DEVELOPMENTS IN
EARTH & ENVIRONMENTAL SCIENCES 3

Oil Pollution and its Environmental Impact in the Arabian Gulf Region

DEVELOPMENTS IN **EARTH & ENVIRONMENTAL SCIENCES** **3**

VOLUME 1 GEOSCIENCES, ENVIRONMENT AND MAN
By H. Chamley

VOLUME 2 MEDICAL GEOLOGY
EFFECTS OF GEOLOGICAL ENVIRONMENTS ON HUMAN HEALTH
by M.M. Komatina

DEVELOPMENTS IN **EARTH & ENVIRONMENTAL SCIENCES 3**

Oil Pollution and its Environmental Impact in the Arabian Gulf Region

M. AL-AZAB[1], W. EL-SHORBAGY[2] AND S. AL-GHAIS[1]

[1]FACULTY OF SCIENCE, [2]FACULTY OF ENGINEERING, UNITED ARAB EMIRATES UNIVERSITY, AL-AIN, UNITED ARAB EMIRATES

ELSEVIER

Amsterdam – Boston – Heidelberg – London – New York – Oxford
Paris – San Diego – San Francisco – Singapore – Sydney – Tokyo

ELSEVIER B.V.
Radarweg 29
P.O. Box 211, 1000 AE, Amsterdam
The Netherlands

ELSEVIER Inc.
525 B Street, Suite 1900
San Diego, CA 92101-4495
USA

ELSEVIER Ltd
The Boulevard, Langford Lane
Kidlington, Oxford OX5 1GB
UK

ELSEVIER Ltd
84 Theobalds Road
London WC1X 8RR
UK

First edition 2005

Library of Congress Cataloging in Publication Data
A catalog record is available from the Library of Congress.

British Library Cataloguing in Publication Data
A catalogue record is available from the British Library.

Cover photo: "Torrey Canyon" from this book's Introductory Chapter "The International Oil Pollution Compensation Funds" by Richard Briggs.

ISBN: 978-0-444-52060-9
ISBN: 0-444-52060-0
ISSN: 1571-9197

♾ The paper used in this publication meets the requirements of ANSI/NISO Z39.48-1992 (Permanence of Paper). Printed in The Netherlands.

Dedication

We dedicate this publication to late UAE leader
Sheikh Zayed bin Sultan Al-Nahyan,
father of the nation and architect of the modern United Arab Emirates.

……… We cherish our environment because it is an integral part of our country, our history and our heritage. On land and in the sea, our forefathers lived and survived in this environment. They were able to do so only because they recognized the need to conserve it, to take from it only what they needed to live, and to preserve it for succeeding generations. With God's will, we shall continue to work to protect our environment and our wildlife, as did our forefathers before us. It is a duty: and, if we fail, our children, rightly, will reproach us for squandering an essential part of their inheritance, and of our heritage.

(From Sheikh Zayed's speech on the occasion of the UAE's first Environment Day in February 1998).

Contents

List of Contributors

Raeid M.M. Abed Max-Planck Institute for Marine Microbiology, Celsiusstrasse 1, D-28359 Bremen, Germany

Mohammad Yousif Al Jawdar Environmental Research and Wildlife Development Agency, P.O. Box 45553, Abu Dhabi, United Arab Emirates

Saif M. Al-Ghais Biology Department, Faculty of Science, UAE University, P.O. Box 17551, Al-Ain, United Arab Emirates

E. Al-Hogaraty Arab Academy for Science and Technology and Maritime Transport, P.O. Box 52212, Dubai, United Arab Emirates

Mir Azzam Scientific Research Sector, United Arab Emirates University, P.O. Box 17555, Al-Ain, United Arab Emirates

Richard Briggs Al Tamimi & Company, Dubai, United Arab Emirates

Bogdan Catalin Department of Biological Sciences, University of Illinois at Chicago, Chicago, IL 60607, USA

M.A.H. Deshgooni Environment Protection and Safety Section, Dubai Municipality, Dubai, UAE

V. Djurdjevic Faculty of Physics, Belgrade University, Belgrade, Serbia and Montenegro

Alam El-Din Oceanography Department, Faculty of Science, Alexandria University, Alexandria, Egypt

Walid El-Shorbagy Civil & Environmental Engineering Department, United Arab Emirates University, P.O. Box 17555, Al-Ain, United Arab Emirates

Salem Essa Geology Department, Faculty of Science, UAE University, P.O. Box 17551, Al-Ain, UAE

Stefan Grötzschel Max-Planck Institute for Marine Microbiology, Celsiusstrasse 1, D-28359 Bremen, Germany

W. Hamza Biology Department, Faculty of Science, United Arab Emirates University, P.O. Box 17551, Al-Ain, United Arab Emirates

Hussein Harahsheh Global Scan Technologies, P.O. Box 1286, Dubai, UAE

Paul S. Hill Department of Oceanography, Dalhousie University, Halifax, NS, Canada B3H 4J1

Tahir Imran Qureshi Department of Chemical Engineering, NFC-Institute of Engineering and Technological Training, Fertilizer Project, Multan, Pakistan.

Mohamed A. Karim Chemical Engineering Department, United Arab Emirates University, P.O. Box 17555, Al-Ain, United Arab Emirates

Arzhang Khalili Max Planck Institute for Marine Microbiology, Celsiusstr.1, 28359 Bremen, Germany

Yaseen H. Khan Department of Biological Sciences, University of Illinois at Chicago, Chicago, IL 60607, USA

Ali Khelifa Department of Oceanography, Dalhousie University, Halifax, NS, Canada B3H 4J1

Young-Ju Kim Department of Environmental Engineering, Kyungpook National University, Daegu, South Korea

J. Koponen Environmental Impact Assessment Center of Finland Ltd., Tekniikantie 21 B, FIN-0215 Espoo, Finland

Kenneth Lee Center for Offshore Oil and Gas Environmental Research, Bedford Institute of Oceanography, Fisheries and Oceans Canada, P.O. Box 1006, 1 Challenger Drive, Dartmouth, NS, Canada, B2Y 4A2

J.H. Lee Department of Environmental Engineering, Kyungpook National University, Daegu, South Korea

A. Micallef Euro-Mediterranean Centre on Insular Coastal Dynamics (ICoD), Foundation for International Studies, University of Malta, Malta

Hassan Mohammadi ROPME, P.O. Box 26388, Safat 13124, State of Kuwait, Kuwait

S. Music Euro-Mediterranean Centre on Insular Coastal Dynamics (ICoD), Foundation for International Studies, University of Malta, Malta

Shahid Mustafa Environment Protection and Safety Section, Dubai Municipality, Dubai, UAE

S. Nickovic Euro-Mediterranean Centre on Insular Coastal Dynamics (ICoD), Foundation for International Studies, University of Malta, Malta

Takashi Nishidai JGI, Inc., 1-5-21, Otsuka, Bunkyo-ku, Tokyo 112-0012, Japan

Takumi Onuma JGI, Inc., 1-5-21, Otsuka, Bunkyo-ku, Tokyo 112-0012, Japan

M.A. Quddus Khan Department of Biological Sciences, University of Illinois at Chicago, Chicago, IL 60607, USA

B. Rajkovic Faculty of Physics, Belgrade University, Belgrade, Serbia and Montenegro

J. Sarkkula Finnish Environment Institute (FEI), P.O. Box 140, FIN-00251 Helsinki, Finland

Manabu Shiobara Japan Oil Development Co., Ltd., 1-21-2, Shinkawa, Chuo-ku, Tokyo 104-0033, Japan

B. Telenta Euro-Mediterranean Centre on Insular Coastal Dynamics (ICoD), Foundation for International Studies, University of Malta, Malta

Preface

The Arabian Gulf (referred to in some presentations in this book as the Persian Gulf) is an arm of the Arabian Sea. It represents a stressed ecosystem because it is situated within the richest oil province in the world that hosts more than 67% of the world oil reserve. Oil-related activities that range from exploration to exportation result in a wide range of adverse effects that cause significant damages to the components of the ecosystem such as coral reefs, algal mats, mangrove and other habitats.

An international conference was held in United Arab Emirates during the period of October 5–7, 2003 to exchange ideas and information among interested members of the scientific community regarding the main aspects of oil pollution in the Arabian Gulf region. More than 30 presentations were given during the event by scientists and interested professionals from more than 15 countries. A number of Keynote Talks were also presented by well-known regional and international scholars. The main aspects discussed in the conference included:

1. Identification and documentation of the major sources of oil pollution in the Gulf region.
2. Evaluation of the analytical methods used to identify the different types of pollutants.
3. Review of the recent advances in oil pollution impact treatment and prevention.
4. Develop stronger cooperation ties between interested members of the community.
5. Encourage awareness of the oil pollution as a serious environmental problem in the region.

This book documents the main articles presented during the conference. It is divided into four main sections as follows:

1. Monitoring and Characterizing Oil Spills.
2. Modeling the Fate of Pollutants and Oil Slicks in Marine Water.
3. Environmental Effects of Oil Pollution on the Ecosystem Components.
4. Combating, Prevention and Treatment of Oil Pollution.

The book starts with an article about the International Oil Pollution Compensation Funds. It presents a brief outline of the history and workings of the oil pollution compensation funds, mainly the 1971 and 1992 funds. More details about the 1992 fund are presented.

The second section titled "Monitoring and Characterizing Oil Spills" has six articles. The first two articles deal with the use of satellite images in monitoring coastal environment and detecting oil spill taking place in the United Arab Emirates coastal waters. A number of satellite data are processed and analyzed using images from early US intelligence satellites of the mid-1970s to the most recent earth observation satellites, European Envisat, 2003. The fourth article presents the use of remote sensing in providing synoptic, repetitive and multispectral data to serve the inventory and monitoring of coastal features, such as tidal wet-lands, potential aquaculture sites, mangroves, estuary-dynamics/shoreline-changes, and off-shore aspects like suspended sediment dynamics and coastal currents, near-shore bathymetry, internal waves, etc. The next article describes the atmospheric-marine system RAMSES designed to provide necessary marine and weather predictions useful for oil spill monitoring. The system has been operationally implemented over the Mediterranean. Recent extensions of the system are introduced and a new oil spill model being under development is fully embedded into the ocean model. Elementary results of a hypothetical oil spill release simulation in the Persian Gulf are reported. Another study presents several techniques used in investigating the structure and function of microbial communities in oil biodegradation studies while the last study addresses the role of oil–sediment aggregation in dispersion and biodegradation of spilled oil. Aggregation takes place between suspended sediment grains and crude-oil droplets forming what is referred to as oil–mineral aggregates (OMA). Such process is found to enhance the dispersion of spilled oil and its biodegradation in aquatic environments. The studies elaborate on the mechanism of formation and the effects of environmental parameters on such formation such as salinity, sediment concentration and size as well as oil type.

Section III titled "Modeling the Fate of Pollutants and Oil Slicks in Marine Water" has four studies. The first two studies present different numerical techniques in modeling the fate transport of pollutants in coastal water. Both studies use three-dimensional formulation; the first is applied in the Abu-Qir Bay in East Alexandria, Egypt while the second is applied in Ruwais area of United Arab Emirates. The third study presents a recently upgraded numerical model (MCOMA2) that investigates the minimum sediment concentration for OMA formation discussed earlier in Section I. Minimum sediment concentration is defined as the concentration of sediment below which concentration of OMA-stabilized droplets is negligible (less than 1%). Accommodating and controlling such aggregation phenomena in real applications is discussed to show its role in enhancing the natural recovery of oiled shorelines when the method of "surf washing" is applied. The last study in this section introduces a mathematical method to quantify interfacial fluxes near and through preamble interfaces, simultaneously, and without using hydrodynamic interface conditions that can be applied for calculation of contaminant transport through permeable sediments.

Section IV titled as "Environmental Effects of Oil Pollution on the Ecosystem Components" contains two studies addressing direct effect of oil pollution and spilled hydrocarbons on various marine fauna and flora. The lower vertebrates are assed in the first paper, while the lower trophic organisms; mainly phytoplankton and zooplankton are investigated in the second paper.

The last section titled as "Combating, Prevention and Treatment of Oil Pollution" has two papers. The first paper presents an overall discussion of preventing and/or combating oil pollution in the ROPME Sea water. ROPME stands for Regional Organization for the Protection of the Environment and it includes all the countries overlooking the Arabian

Gulf. The last paper presents a treatment technique to the oily wastewater by the means of UV-Catalytic technology.

The Editors
M. Al-Azab, W. El-Shorbagy & S. Al-Ghais
January 2005

Acknowledgment

We would like to thank His Highness Sheikh Nahyan Mubarak Al Nahyan, Minister of Education and Chancellor of the United Arab Emirates University for his inspiration, encouragement and support.

The editors would like to thank all who contributed to the conference from which this book was originated as well as to those who helped in the book production. The acknowledgment is due to the members of the Conference Organizing Committee that included:

- Professor Joseph Hill, Dean, Faculty of Science (Chairman)
- Dr. Mohamed Al-Azab (Vice-Dean, Faculty of Science)
- Dr. Abdulla K. Musallam (Asst. Dean of Student Affairs, Faculty of Science)
- Dr. Wajdi Zoghaib (Chemistry Dept., Faculty of Science)
- Dr. Saif Al-Ghais (Biology Dept., Faculty of Science)
- Dr. Waleed Hamza (Biology Dept., Faculty of Science)
- Dr. Esam Abd El-Gawad (Geology Dept., Faculty of Science)
- Dr. Fadhil N. Sadooni (Geology Dept., Faculty of Science)
- Professor Rachid Chebbi (Chemical and Petroleum Engineering Dept., Faculty of Science)
- Dr. Walid El-Shorbagy (Civil and Environmental Engineering Dept., Faculty of Engineering)
- Mr. Mahfouz Abdullah (Public Relations)
- Mr. Adel Usman (Finance)

Special thanks to Dr. Esam Abd El-Gawad and Professor Rachid Chebbi who contributed with the editors in reviewing the abstracts. Acknowledgment is due to Mr. Mohamed Shahid and Mr. Hamdi Kandil from the Faculty of Science who has made a great effort in editing the book.

Finally, we thank Elsevier for their patience and encouragement from the inception of this book to its completion.

ADNOC

Established in 1971, Abu Dhabi National Oil Company (ADNOC) is one of the world's leading companies with major oil reserves and in recent years, exploration activities have continued using state-of-the-art seismic analysis in order to increase its proven reserves.

Towards achieving its goal to make ADNOC a world-class environmental company, the focus of ADNOC and its Group of Companies on health, safety and environment has been growing in importance and includes greater protection to the population and country at large, on land and sea against industrial pollution. The achievements include dramatic reduction of gas flaring. It is ADNOC's ultimate objective to completely eliminate gas flaring.

I

Introduction

Oil Pollution and its Environmental Impact in the Arabian Gulf Region
M. Al-Azab, W. El-Shorbagy and S. Al-Ghais, editors

Chapter 1
The international oil pollution compensation funds

Richard Briggs
Al Tamimi & Company, Dubai, United Arab Emirates

Abstract

Starting with the Torrey Canyon incident in 1967, compensation for oil pollution generated from shipping tankers has evolved in many shapes till that day. This article presents a brief outline of the history and workings of the oil pollution compensation funds. Two main funds are discussed; the 1971 and 1992 Funds. The legal principles of the 1992 Fund are further detailed out and the claims handling procedures are outlined. The article concludes with addressing the future of the 1992 Fund, its challenges and opportunities.

1. Background

Oil Pollution Compensation has come a long way since the "Torrey Canyon" incident in 1967. It was this incident, which introduced the public at large to issues of oil pollution, and created the first real public outcry on this subject. This in turn provided the stimulus to develop international conventions through which compensation would be available to those who incur clean up costs or suffer pollution damage as a result of a spill of persistent hydrocarbon mineral oil from a tanker. The term "persistent oil" can be taken to include crude oil, heavy and medium fuel oil, heavy diesel oil and lubricating oil.

Voluntary agreements such as the Tanker Owners Voluntary Agreement Concerning Liability for Oil Pollution (Tovalop) and the Contract Regarding a Supplement to Tanker Liability for Oil Pollution (Crystal) were established by the tanker and oil industries in the 1960s as interim arrangements pending the widespread ratification of two International Conventions. As countries around the world ratified the subsequent Conventions, and pursuant to the entry into force in 1996 of protocols which updated the original Conventions, the voluntary agreements of Tovalop and Crystal were terminated on 20 February 1997.

The conventions to deal with the consequences of oil pollution were developed under the auspices of the International Maritime Organisation (IMO). The original conventions were the 1969 International Convention on Civil Liability for Oil Pollution Damage (1969 CLC) and the 1971 International Convention on the Establishment of an International

E-mail address: r.briggs@tamimi.com (R. Briggs).

Figure 1. The Torrey Canyon, 1967.

Fund for Compensation for Oil Pollution Damage (1971 Fund Convention). This "old" regime was amended in 1992 by two protocols which increased the compensation limit and broaden the scope of the original conventions. The amended conventions are usually known as the 1992 Civil Liability Convention (1992 CLC) and the 1992 Fund Convention.

2. Introduction to the 1971 and 1992 funds

The International Oil Pollution Compensation Funds 1971 and 1992 (1971 and 1992 Funds) are inter-governmental organizations providing compensation for oil pollution damage resulting from spills of persistent oil from tankers. The International Oil Pollution Compensation Fund 1971, otherwise known as the 1971 Fund, was established in October 1978. It operated within the framework of two international conventions, the 1969 Civil Liability Convention and the 1971 Fund Convention. This "old regime" was amended in 1992 by the two protocols known as the 1992 Civil Liability Convention and the 1992 Fund Convention. The International Oil Pollution Compensation Fund 1992 (1992 Fund) was set up under the 1992 Fund Convention when the latter entered into force.

The 1971 Fund Convention ceased to be in force on 24 May 2002 and is now in "run off". The 1971 Fund Convention, therefore, does not apply to incidents occurring after that date. It was actually the denunciation of the 1971 Fund Convention by the United Arab Emirates, where I practice, which ceased the activities of the 1971 Fund. Incidentally, one of the ongoing claims keeping the 1971 Fund in "run off" is the "Pontoon 300" claim in Umm Al Quwain, UAE. As a result of the "run off" of the 1971 Fund and with an eye to the future, I will only be talking about the 1992 Fund from this point forward in my paper.

The International Oil Pollution Compensation Fund 1992 ("The 1992 Fund") entered into force on 30 May 1996 for nine States, and by the end of 2001, 62 States had become members of the 1992 Fund. Its purpose is to compensate those suffering from oil pollution damage. The 1992 Fund and provides compensation for oil pollution damage resulting from spills of oil from tankers in States which are members of the organisation. The Maximum compensation payable by the 1992 Fund for any one incident is 135 million SDR (Special Drawing Rights), which is approx £117 million or USD 170 million.

The 1992 Fund is financed by levies on certain types of oil carried by sea, and such levies are paid by entities (companies) which receive oil after sea transport. With regard to the basis for such levies, the 1992 Fund is financed by contributions paid by any person who has received in the relevant calendar year in excess of 150,000 metric tons of crude oil or heavy fuel oil (contributing oil) in ports or terminal insulations in a state which is a member of the 1992 Fund, after carriage by sea. The levy of contributions is based on reports on oil receipts in respect of individual contributors, which are submitted to the Secretariat of the 1992 Fund by the Government of Member States. Contributions are paid by the individual contributors directly to the 1992 Fund. Governments are not responsible for these payments unless they have voluntarily accepted such responsibility. This means that there are some states, the United Arab Emirates and other Gulf Co-operation Council ("GCC") States for example, which export oil but have little to no imports of oil, and therefore receive the benefit of the 1992 Fund without paying much for the same. Japan, on the other hand paid 21% of general fund contributions to the 1992 Fund in 2001, with Italy paying 11%, the Republic of Korea paying 10%, the Netherlands 8%, France 8% and the UK 6% Singapore, relevantly, paid 5% of the 1992 Fund's contribution for the year 2001.

Some states have opted out of the International Oil Pollution Compensation Funds system to rely on their own national laws, or in the case of the United States of America, specifically created their own system to deal with oil pollution, liability and compensation. (US Oil Pollution Act (OPA) 1990). Other States, such as Pakistan with the recent "Tasman Spirit" spill at Clifton Beach, Karachi may come to regret not becoming a party to the Civil Liability and Fund Conventions.

In terms of its set up, the 1992 Fund has an Assembly composed all Member States of the organization, as well as an Executive Committee of 15 member states elected by the Assembly. The main function of such Executive Committee is to approve settlements of claims for compensation to the extent that the 1992 Fund's Director is not authorised to make such settlements. The 1992 Fund is administered by a Secretariat, headed by a Director. The Secretariat is located in London, and the current Director is Mr. Mans Jacobsson. The Director is authorised to make settlements and pay compensation if it is unlikely that the total payments in respect of the incident will exceed SDR 2.5 million (approximately USD 3.5 million) without any specific authority from the 1992 Fund's Executive Committee.

3. Legal principles of the 1992 fund

As stated above, the 1992 Fund is able to pay compensation for oil damage only in respect of claims fulfilling the criteria for admissibility laid out in the relevant international conventions, namely 1992 Civil Liability Convention and the 1992 Fund Convention.

The 1992 Civil Liability Convention covers pollution damage suffered in the territory or territorial sea or exclusive economic zone of a State which is a party to the Convention. The UAE and Singapore (as examples) are parties to the 1992 Civil Liability Convention and 1992 Fund Convention, and therefore receive the benefits of the same. The 1992 Civil Liability Convention and 1992 Fund Convention cover spills of persistent crude and fuel (bunker) oil from sea going tankers, and can apply to both laden and unladen tankers (but not dry cargo ships).

I have set out below a chart which sets out in broad terms how the "two tier" compensation system works

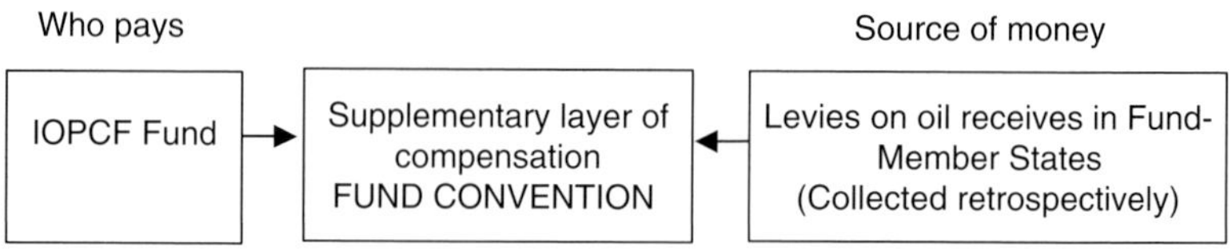

The 1992 Civil Liability Convention may be understood as a "two tier" system of compensation and is based on the principal of "strict liability," which means that the owner of the tanker which spills the oil is liable regardless of whether or not he was actually at fault, subject to certain exemptions such as "force majeure," act of war or sabotage by a third party, or negligence by public authorities. The idea is that the claimants (who have suffered loses from the oil spill) can receive relatively prompt compensation without the need for lengthy and costly litigation, and the ship owner can deal with the liabilities of an oil spill via a known and established system which attempts to standardise legal and practical issues in different jurisdictions.

Under the 1992 Civil Liability Convention, the tanker owner which spilled the oil will normally be entitled to limit liability to an amount based on the gross tonnage of the tanker involved in the incident. If it is proved, however to the satisfaction of the court in question that any pollution damage resulted from the owners personal act or omission, or was done with the intent to cause pollution damage, or recklessly with the knowledge that such damage will probably result, the owner will be deprived of the right to limit liability.

The registered owner of the tanker concerned or his pollution liability insurer (usually a Protection and Indemnity Association – "P&I Club") bear the first part of the combined value of the claims up to the tanker owner's limitation of liability figure. Further to that, the 1992 Fund may provide supplementary compensation where the compensation available from the tanker owner under the 1992 Civil Liability Convention is insufficient to meet all valid claims. In some cases, the 1992 Fund may meet the totality of claims compensation if, for example, a tanker owner is uninsured and insolvent. If the State where the spill occurred is the member of the 1992 Fund and the oil spill originated from a tanker as defined in 1992 Civil Liability Convention, the 1992 Fund will pay, provided claims are correctly substantiated. If the total of all respective claims for pollution damage exceeds the total amount of compensation available under the 1992 Civil Liability Convention and 1992 Fund Convention, the compensation paid to each claimant will be reduced proportionally on a "pro rata" basis. This can create problems for the 1992 Fund, particularly as the Fund can be forced to make "interim payments" in circumstances where there are disputed claims outstanding and it cannot be ascertained what the final level of claims will be. This is currently the case in relation to claims against the 1971 Fund in Umm Al Quwain on the "Pontoon 300" incident.

The 1992 Fund have the responsibility of administrating the regime of compensation created by the 1992 Fund Convention. The 1992 Fund pays supplementary compensation, which may be available when the compensation available from the tanker owner under the 1992 CLC is insufficient to meet all valid claims. In rare cases, and we have seen this in the UAE in relation to uninsured barges carrying oil, the 1992 Fund (or its predecessor, the 1971 Fund) may meet the totality of claims in compensation if, for example, the tanker owner cannot be identified, or if the tanker owner is uninsured and insolvent, or if the tanker owner is exonerated from the liability under certain provisions in the 1992 CLC which do not apply in the case of the 1992 Fund Convention.

The Fund will also not pay compensation if the damage occurred in a state which was not a member of the 1992 Fund, or if the pollution damage resulted from an act of war, or was caused by spill from a war ship.

4. Claims handling procedure

Any claims for compensation under the 1992 CLC should be brought against the tanker owner or directly against his P&I insurer. To obtain compensation from the 1992 Fund, any claimant should submit his claim directly to the Secretariat of the 1992 Fund. The 1992 Fund takes an active interest in any incident where it appears that the 1992 Fund may be called upon to pay compensation. The P&I Club and 1992 Fund will often jointly investigate and assess the damage from a particular incident, and will co-operate closely in the settlement of claims to ensure a consistent and efficient approach. In some cases, this can lead to joint claims offices with the P&I Club's local correspondent and local surveyor. In particularly difficult cases of large spills such as "Erika" and "Prestige," offices in the locality of the spill may be set up specifically by the 1992 Fund to deal with such single incidents.

Technical co-operation between P&I Clubs and the 1992 Fund is standard in the event of a spill, and this often extends to the appointment of the same technical advisers and experts. In most cases, a member of the technical staff of the International Tanker Owners Pollution Federation (ITOPF) will be asked to attend on-site at the tanker spill site by both the P&I Club and the 1992 Fund. ITOPF staff has substantial practical experience of combating marine oil spills, having attended at over 400 incidents in various different countries including the UAE. ITOPF is often involved in the post-spill assessment of the technical merits of claim for clean up costs and damage arising from cases attended on site. However, any final decision on the admissibility of claims and the appropriate settlement levels rests solely with the P&I Club concerned and the 1992 Fund.

For a claim to be admissible, it must fall within the definition of pollution damage or preventive measures in the 1992 Civil Liability Convention and 1992 Fund Convention. A uniform interpretation of the definitions set out in the Conventions and a common understanding of what constitutes an admissible claim are essential for the efficient functioning of the International system for compensation established by the conventions, and the 1992 Fund's Executive Committee and Secretariat places particular stress on the importance of a uniform interpretation of the Conventions by the courts of the States involved. The 1992 Fund has a claims manual in respect of "claims handling" which is readily available.

Broadly speaking, claims in respect of pollution damage are under four categories:

1. Preventive measures (including clean-up);
2. Damage to property;
3. Economic losses; and
4. Reinstatement/restoration of impaired environments.

Anyone who has suffered pollution damage (including the taking of preventative measures) in a State which is a party to the 1992 Civil Liability Convention and/or 1992 Fund Convention may claim for compensation. Claimants may be private individuals, partnerships, companies (including ship owners, charterers and terminal operators) or public bodies (including central and local government authorities and agencies). Claims are subject to time limits, and claimants will lose the right to compensation unless a court action is brought against the tanker owner and his P&I Club or against the 1992 Fund within three years of the date on which the damage occurred. Although damage may occur some time after an accident takes place, court proceedings must in all cases be brought within six years of the date of any incident.

The 1992 Fund, and its predecessor the 1971 Fund, have been involved in numerous oil pollution incidents, including the "Braer" off Scotland (1993), "Sea Empress" off Wales (1996), "Nakhodka" of Japan (1997), "Nissos Amorgos" off Venezuela (1997) and "Erika" off France (1999), "Prestige" off Spain (2002) as well as UAE related incidents, "Pontoon 300" off Umm Al Quwain (1998), "Al Jaziah 1" off Abu Dhabi (2000) and "Zeinab" off Dubai (2001).

5. Conclusion – the future

5.1. What is the future for the 1992 Fund?

In its expected by the Secretariat that there will be considerable growth in the 1992 Funds membership in the next couple of years, as remaining 1971 Fund member States ratify the 1992 Fund Convention and more States which were not previously members of the 1971 Fund join the 1992 Fund. The latest State to sign up is Cape Verde, which will join the 1992 Fund with effect from 04 July 2004. There is the usual set of new spills which the 1992 Fund is involved with, the best known being the sinking of the tanker "Prestige" in international waters off north–west Spain at the end of last year. The stated policy of the 1992 Fund remains to settle claims arising out of each incident as quickly as possible and the 1992 Fund remains under considerable pressure in the event of a spill to react to the presentation of claims in a prompt and organised fashion. This pressure will no doubt increase in coming years.

The 1992 Fund also set up a Working Group to consider the need to improve the international compensation regime. The Working Group had Mr. Alfred Popp Q.C. as its Chairman, and since its first meeting held in June 2000 exchanged views concerning the need for and possibilities of improving the compensation regime established by 1992 Civil Liability Convention and 1992 Fund Convention. It also considered issues regarding the maximum levels of compensation, ship-owners liability, environmental damage and environmental studies. It discussed the admissibility of claims for fixed costs, time-bar, alternative dispute settlement procedure, and the ongoing problems caused by some States non-submission of oil reports (used to calculate levies charged by the 1992 Fund). Such Working Group also prepared a draft protocol establishing a "Supplementary Fund" for Compensation, which was adopted at a diplomatic conference held at the Head Quarters of the IMO in London in May 2003. This "Supplementary Fund" allows additional compensation to be available for future victims of pollution from oil tanker accidents.

The aim of the "Supplementary Fund" is to supplement compensation available under the 1992 Civil Liability and Fund Conventions with an additional third tier of compensation. Membership of the "Supplementary Fund" is optional and any State which is a Member of the 1992 Fund may join the "Supplementary Fund".

The total amount of compensation payable for any one incident will be 750 Million Special Drawing Rights ("SDR's") which in cash terms, equates to just over US$1,000,000,000 (1 billion) per incident. This includes the amounts payable under the existing 1992 Civil Liability Convention and Fund Convention.

The "Supplementary Fund" Protocol will enter into force three months after it has been ratified by at least eight States which have received a combined total of 450 Million tonnes of contributing oil in a calendar year. The "Supplementary Fund" will only pay compensation for pollution damage in States which are members of the "Supplementary Fund" for incidents which occur after the protocol has entered into force.

II
Monitoring and characterization oil spills

Oil Pollution and its Environmental Impact in the Arabian Gulf Region
M. Al-Azab, W. El-Shorbagy and S. Al-Ghais, editors

Chapter 2

Monitoring of coastal environment using satellite images in the United Arab Emirates

Mohammad Yousif Al Jawdar[1],
Manabu Shiobara[2] and Takumi Onuma[3,*]

[1]*Environmental Research and Wildlife Development Agency, P.O. Box 45553, Abu Dhabi, United Arab Emirates*
[2]*Japan Oil Development Co., Ltd., 1-21-2, Shinkawa, Chuo-ku, Tokyo 104-0033, Japan*
[3]*JGI, Inc., 1-5-21, Otsuka, Bunkyo-ku, Tokyo 112-0012, Japan*

Abstract

The diversity of marine wildlife along the coast of the United Arab Emirates is well known as well as the offshore giant oilfields. The protection of sustainable marine environment is the first priority to keep the biodiversity in the region and the Environmental Research and Wildlife Development Agency (ERWDA) has been paying attention on the preservation of coastal environment since it was established in 1996. It is widely accepted that long-term monitoring is mandatory for environmental hazard and satellite images with intermediate spatial resolution covering the broad region are suitable for this purpose. This study is based on such a viewpoint to contribute to the monitoring of environmental change along the coast of Abu Dhabi Emirate. The methodology of producing bottom type maps and vegetation distribution maps from satellite image data are presented. The operational atmospheric correction algorithm to minimize air–water interface was designed and applied to Terra Advanced Spaceborne Thermal Emission and Reflection (ASTER radiometer) data to produce Bottom Index images, which were compared with the field survey results to convert the index value to the bottom type. Bottom index value of 2.5 was deemed to discriminate sea grass and seaweed from sand bottom. The distribution of onshore vegetation was extracted by using Soil Adjusted Vegetation Index, one of the indices dealing with vegetative activity, and then was classified into two vegetation types by the supervised classification method using [Natural Environmental Resources and their Oil Spill Protection Priorities for the Coastline of Abu Dhabi Emirate, U.A.E, p. 120, 2000] as the supervisor.

1. Introduction

There are well preserved coral banks, reefs and shallow lagoons along the long coastline of Abu Dhabi Emirate in the southern Arabian Gulf. The coastal area is characterized by the biodiversity of marine wildlife, including endangered species of dugongs and turtles.

---

*Corresponding author. *E-mail address:* onuma@jgi.co.jp (T. Onuma).

Mangrove forests and algal mats develop naturally along the tidal limits of the lagoons. The conservation of the biodiversity is major concern on the protection of coastal environment in the Emirate.

Oil pollution related with the offshore petroleum development, such as leakage from production facilities and pipelines, oil tanker accidents, etc. obviously has severe impact on the coastal environment and is the issue on conservation of the ecosystem in coastal regions. In order to respond to oil spill events the Abu Dhabi Oil Spill Protection Priorities Maps 2000 (ERWDA, 2000) was prepared by using Landsat 5 TM images acquired in 1998 with the spatial resolution of 30 m. Recent satellite images, however, show clearly that dredged canals and newly constructed facilities have changed coastal feature since then. Thus updating of the maps using new satellite images with finer spatial resolution is urgently needed for the protection of coastal environment.

The study aims to develop the operational algorithms for producing bottom type maps in the shallow waters and vegetation maps of the coastal regions by using Terra Advanced Spaceborne Thermal Emission and Reflection (ASTER radiometer) images. The shallow sea with water depth less than 5 m, where Terra ASTER can detect the reflection from the sea floor, covers broad intertidal areas in the Abu Dhabi Emirate. Imaging the detailed distribution of bottom types by using Terra ASTER data with the spatial resolution of 15 m is one of the goals of this study. It is well known that the vegetation of the intertidal areas plays a significant role in the ecosystem of coastal regions. ERWDA and oil development companies like Japan Oil Development Co., Ltd. (JODCO) have contributed to mangrove plantation for the conservation of the coastal environment. The study also aims to extract vegetation distribution in the coastal areas from Terra ASTER images, as well as to classify vegetation types.

The details of algorithms are described first and followed by the examples which were analyzed from Terra ASTER images and the field verification conducted by ERWDA.

2. Satellite image data

Twenty scenes of Terra ASTER data acquired from 2000 through 2003 are used in the study. ASTER is a cooperative effort between NASA and Japan's Ministry of Economy, Trade and Industry, and consists of three instrument subsystems referred as Visible-Near InfraRed (VNIR), ShortWave InfraRed (SWIR) and Thermal InfraRed (TIR). Each subsystem operates in a different spectral region, and has its own telescope. In this study VNIR data is mainly used because green (Band 1) and red (Band 2) can reach the sea bottom down to the depths of 15 and 5 m, respectively, while the energy at Bands 3 through 14 is completely absorbed. As chlorophyll absorption is an important input for vegetation models and a key factor in many vegetation indices, infrared (Band 3) and red bands are frequently used for vegetation analysis. SWIR and TIR data are also used to classify vegetation and to detect clouds. ASTER also has a pointing capability enabling the revisit of 5 days despite of swath width of 60 km that is narrower than the orbital spacing of 172 km. Because of this capability, the geometry of the sun–sea surface–sensor must be considered for precise measurement of the reflectance of surface object and so-called water-leaving radiance.

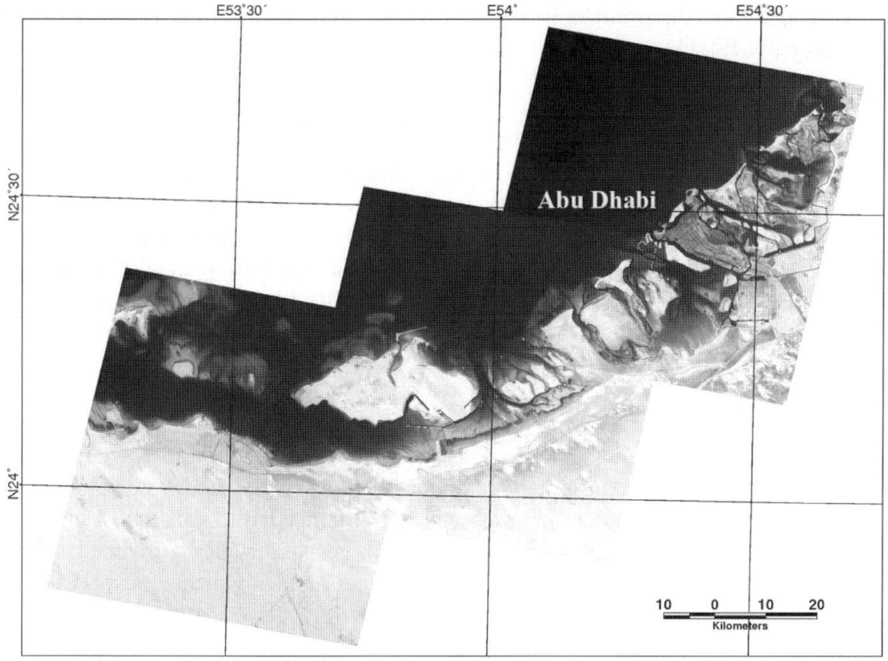

Figure 1. Terra ASTER VNIR mosaic image covering coastal region of Abu Dhabi Emirate. © METI/NASA.

Northeast to the Abu Dhabi Island was selected as a "pilot" area (Fig. 1) for verifying algorithms to be applied. Several scenes were used to cover the area based on time span for surface monitoring and data quality including acquisition condition.

3. Methodology

3.1. Water-leaving radiance

The signal received over the ocean surface consists of atmospheric backscattering components and water-leaving radiance. The radiance backscattered from the atmosphere and/or the sea surface has typically at least an order of magnitude larger than the desired radiance scattered out of the water. Therefore, the process of retrieving water-leaving radiance from the total radiance is needed. According to Gordon et al. (1983), the sensor radiance Lt(λ) can be divided into its components:

$$\mathrm{Lt}(\lambda) = \mathrm{Lr}(\lambda) + \mathrm{La}(\lambda) + t(\lambda)\mathrm{Lw}(\lambda) \quad (1)$$

where Lr(λ) is the contribution arising from Rayleigh scattering, La(λ) is the contribution arising from aerosol scattering, Lw(λ) is the water-leaving radiance, and $t(\lambda)$ is the diffuse transmittance of the atmosphere between sea surface and the sensor. This approximation is assumed for application to CZCS data that has the tilting capability to avoid sun glint.

In order to apply the algorithm to Terra ASTER data, a component of sun glint Lg(λ) must be accounted in Eq. (1):

$$\mathrm{Lt}(\lambda) = \mathrm{Lr}(\lambda) + \mathrm{La}(\lambda) + \mathrm{Lg}(\lambda) + t(\lambda)\mathrm{Lw}(\lambda) \quad (2)$$

Lr(λ) is a function of wavelength, Rayleigh optical thickness and ozone optical thickness. Rayleigh optical thickness is also a function of wavelength. Ozone optical thickness is calculated by using the amount of ozone. In the study, monthly average of TOMS data were used for the amount of ozone. $\tau(\lambda)$ is the function of Rayleigh optical thickness, ozone optical thickness and zenith angle of a vector from a pixel to the sensor. Details of the derivation of terms in Eq. (1) are introduced by Gordon et al. (1983).

As for the La(λ), assuming the Lw(IR) (infrared) is equal to zero due to the absorption by the ocean, then from Eq. (2);

$$\mathrm{La(IR)} = \mathrm{Lt(IR)} - \mathrm{Lr(IR)} - \mathrm{Lg(IR)}$$

Lt(IR) is satellite data and Lr(IR) is known. Lg(IR) is a function of Rayleigh optical thickness, ozone optical thickness, aerosol optical thickness τ_a(IR) and wind velocity. That is, Lg(IR) is needed to calculate La(IR), and τ_a(IR) is needed to calculate Lg(IR). But La(IR) must be known in order to calculate τ_a(IR), because τ_a(IR) is inversely derived from La(IR). To solve this problem, the optical thickness data of typical aerosol (Elterman, 1968) was referred in our algorithm as the start value of iterative search. For IR band, the algorithm first calculates Lg(IR) using the start values of aerosol optical thickness for infrared band and derives Lt(IR) − Lr(IR) − Lg(IR). Then it calculates new τ_a(IR) and recalculates Lg(IR) using this τ_a(IR). Repeat the calculation until the difference of first τ_a and new τ_a becomes less than 0.05. When the condition met, take it as the aerosol optical thickness of the pixel and calculate Lg(IR) as the final.

Scattering phase function by Haltrin (1998) for aerosol of maritime atmosphere was used in the study to calculate τ_a(IR). The single scattering albedo of aerosol for the seawater is stable for one Terra ASTER scene and can be considered as a constant. The result of simulation shows about 0.9 around the Arabian Peninsula (Takemura and Nakajima, 2002).

Assuming that the aerosol optical thickness follows:

$$\tau_i = \alpha\lambda_i^{-\beta}$$

where τ_i and λ_i are the aerosol optical thickness and the central wavelength, respectively, for band i; $i = 1, 2, 3$ for ASTER case. β is so-called the angstrom index. By taking ratio of the aerosol optical thickness between IR band and another band,

$$\tau_a(\lambda) = \tau_a(\mathrm{IR})(\lambda_{\mathrm{IR}}/\lambda)^{\beta}$$

In most studies λ^{-1} function is assumed for the spectral variation of the aerosol optical thickness (Gordon et al., 1983; Ouaidrari and Vermote, 1999; Zhang et al., 1999), in which β is 1. In the ASTER case, therefore,

$$\tau_a(\lambda) = \tau_a(\mathrm{VNIR3})\, 0.807/\lambda$$

where λ is the center wavelengths (in μm) of ASTER VNIR 1 and 2 (= 0.556 and 0.659 μm). Then, τ_a(VNIR1) and τ_a(VNIR2) are derived from τ_a(VNIR3), and La(VNIR1), Lg(VNIR1), La(VNIR2) and Lg(VNIR2) can be calculated from

τ_a(VNIR1) and τ_a(VNIR2), finally the water-leaving radiance is led from the Eq. (2) as

$$\mathrm{Lw}(\lambda_i) = \{\mathrm{Lt}(\lambda_i) - \mathrm{Lr}(\lambda_i) - \mathrm{Lg}(\lambda_i) - \mathrm{La}(\lambda_i)\}/t(\lambda_i)$$

The accuracy of Lw depends on that of $\tau_a(IR)$. Lw sometimes becomes negative due to overestimation of $\tau_a(IR)$, which is resulted from specification of inappropriate wind velocity that is unknown parameter in the algorithm. The algorithm takes an offset value to adjust the negative value. Under the non-turbid clear water condition, the lights with wavelengths of ASTER VNIR 1 and 2 can penetrate down to the depth of 15 and 5 m, respectively. The offset value should be chosen for the open sea that is adequately deep.

3.2. Bottom index

As for the shallow water of the same bottom type, the relationship between water-leaving radiance of two bands in VNIR region can be expressed as

$$\ln(\mathrm{Lw}_i) = (k_i/k_j)\ln(\mathrm{Lw}_j) + a \qquad (3)$$

where Lw is water-leaving radiance, k is an attenuation coefficient, a is a constant, i and j are band numbers (Lyzenga, 1981). The attenuation coefficient is independent on bottom types and the ratio (k_i/k_j) is constant. Then, the magnitude of the constant a depends on bottom types, and it is called Bottom Index (BI). From Eq. (3) BI is expressed as

$$\mathrm{BI} = \ln(\mathrm{Lw}_i) - (k_i/k_j)\ln(\mathrm{Lw}_j)$$

In the case of Terra ASTER, i and j are 1 and 2, respectively. The ratio of attenuation coefficients was estimated as follows. For many small water areas with the same substratum (sand) at different depths by visual inspection (as shown in Fig. 2), water-leaving radiances for bands 1 and 2 are calculated as described in the Section 3.1. Average values of (log-transformed) water-leaving radiance for bands 1 and 2 are cross-plotted, then the ratio of attenuation coefficients is derived as the slope of the regressed line.

Terra ASTER data of the same area but with different acquisition dates were tested in the study, and it was found that the slopes were almost identical (0.44) and could be treated as a constant (Fig. 3). Therefore, BI is expressed using Terra ASTER for the shallow waters of Abu Dhabi as

$$\mathrm{BI} = \ln(\mathrm{Lw}_1) - 0.44\ \ln(\mathrm{Lw}_2)$$

Figure 4 shows an example of bottom index produced from Terra ASTER data acquired on November 18, 2000. It can be said that the light of ASTER VNIR band 2 penetrates down to the water depth of 5 m, and the processing result is quite concordant with the iso-depth line of 5 m (blue line in Figure 4).

3.3. Vegetation index

Vegetation index is a parameter widely adopted in vegetation researches on remotely sensed data, and many kinds of indices have been proposed for different purposes. Soil Adjusted Vegetation Index (SAVI) is thought to be suitable for the areas where vegetation

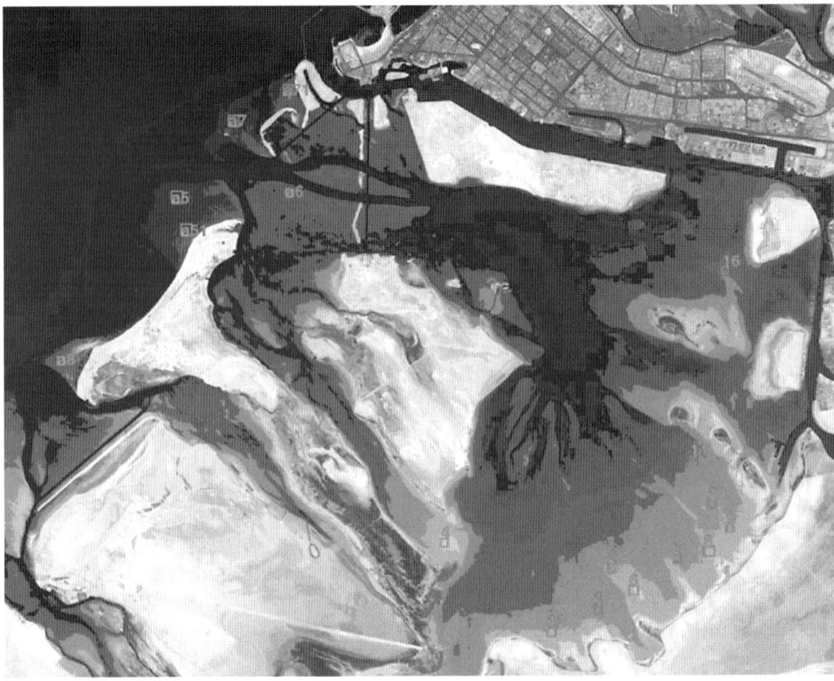

Figure 2. Example of selection of sand bottom at different water depths.

distribution is sparse. As all the indices including SAVI are calculated from reflectance values of infrared and red bands, the atmospheric correction and the conversion from Digital Number (DN) to reflectance of ASTER data must be done first. In this study, the apparent reflectance at the sensor was used instead of one at the ground, because full atmospheric correction of Terra ASTER data using radiation transfer codes is quite time consuming and hence is not operational. The Improved Dark Object Method (Chavez, 1988) was adopted for the atmospheric correction. The method removes the atmospheric scattering based on the image data itself, and the result of the process is given in the

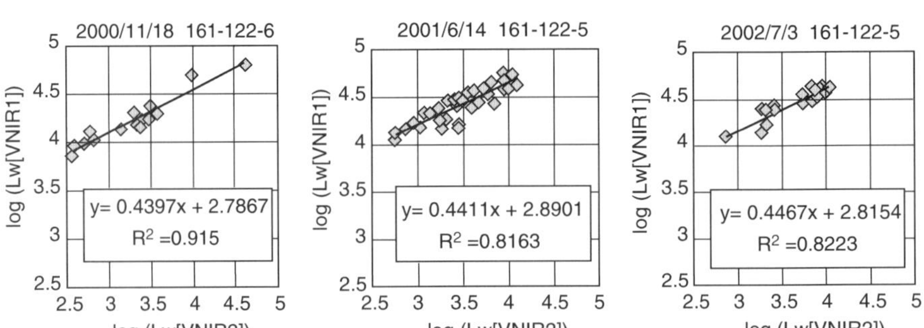

Figure 3. Ratios of attenuation coefficients between ASTER VNIR bands 1 and 2.

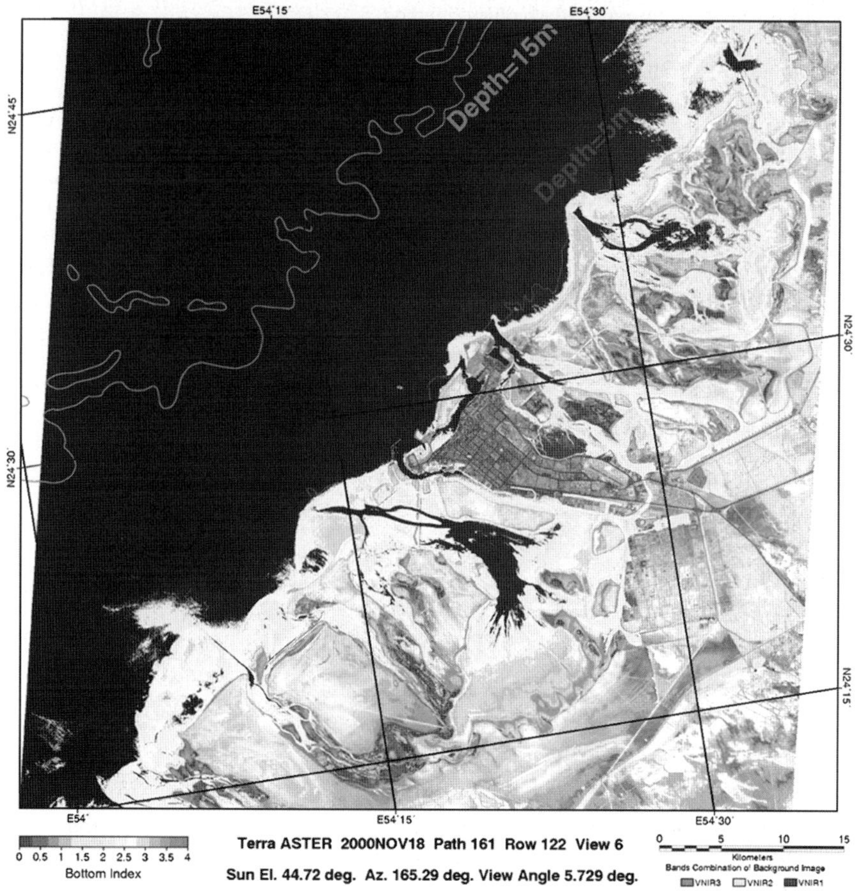

Figure 4. Bottom index image superimposed on Terra ASTER VNIR false color composite.

unit of DN. In order to calculate SAVI, therefore, the result of atmospheric correction must be converted into the apparent reflectance as follows:

$$L_i = (\mathrm{DN}_i - H_i)G_i$$

$$\rho_i = \frac{\pi L_i d^2}{E_i \cos\left(\frac{\pi}{2} - \theta\right)}$$

where i is the band number, DN_i is the digital count of raw data, H_i is the offset calculated by the Improved Dark Object Method, G_i is the conversion coefficient for ASTER band i, E_i is exoatmospheric solar irradiance, d is the astronomical distance between the sun and the earth, and θ is the sun elevation angle. Parameters for Improved Dark Object Method are shown in Table 1. The starting haze value (SHV) is calculated by subtracting the offset value of Table 1 (always 1 in the case of ASTER) from the minimum value of VNIR band 1.

Table 1. Parameters for improved dark object method.

Aster Band	Normalized Gain W/m × m × sr × μm (NORM)	Offset	Irradiance	Normalized Multiplication Factors (MLF)				
				Very clear $1/\lambda^4$	Clear $1/\lambda^2$	moderate $1/\lambda^1$	hazy $1/\lambda^{0.7}$	very hazy $1/\lambda^{0.5}$
1	1.00	1.00	1843.00	1.000	1.000	1.000	1.000	1.000
2	0.95	1.00	1555.00	0.532	0.729	0.854	0.859	0.924
3	0.78	1.00	1111.70	0.286	0.534	0.731	0.803	0.855

Then H_i is given by

$$H_i = \text{SHV} \times \text{MLF}(\text{case}, i) \times \text{NORM}(i) + \text{Offset}(i)$$

and SAVI is given by

$$\text{SAVI} = \frac{(1 + L)(\rho_3 - \rho_2)}{(\rho_2 + \rho_3 + L)}$$

where ρ_2 and ρ_3 are the reflectance of ASTER VNIR bands 2 and 3, respectively, and L is soil adjustment parameter and 0.5 in this study.

4. Bottom type map and vegetation map

A goal of the study is to develop the operational methodology for producing the maps of bottom types and vegetation, aiming at the application in environmental monitoring. Water-leaving radiance, bottom index, apparent reflectance and vegetation index are the parameters needed for the production of these maps, of which derivation algorithms are described in the above. From the viewpoint of verification, the results of data processing should be crosschecked by the ground truth. ERWDA carried out the field survey to collect data of bottom types and vegetation distribution in the pilot area in February 2003. The processing results of satellite image data and the maps has been compared by using Natural Resources Maps (ERWDA, 2000) before the survey to investigate the applicability of the algorithms developed in the study.

4.1. Bottom index to bottom type map

Threshold values of bottom index were estimated through the comparison of the Natural Resources Map and the bottom index calculated from Terra ASTER data. The area to the northeast of Abu Dhabi city was chosen for the pilot area, and regions of sandy bottom, sea grass-shallow and mangrove were digitized separately (Fig. 5). Then, the histograms of the bottom index image (Fig. 6) for sandy bottom and sea grass-shallow areas were compared. By the try and error manner two thresholds of 2.49 and 2.72 were set, and provisional bottom type map was produced (Fig. 7). When compared with the Natural Resources Map,

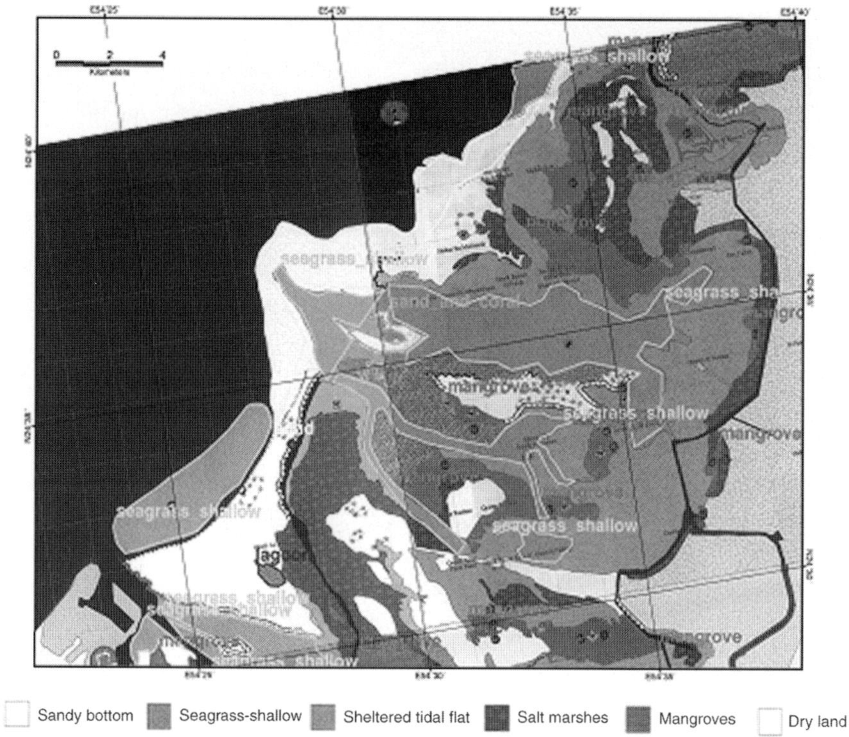

Figure 5. Natural Resources Map (after ERWDA, 2000) and sample areas for comparison with bottom index and vegetation index.

the distribution of the sand in Figure 7 looks quite concordant with that of sandy bottom and sheltered tidal flat in Figure 5. However, because the area does not contain a coral patch and the accuracy of the Natural Resources Map is unknown, the field survey data is needed to clarify the relationship between the ASTER bottom index and the real bottom types.

Figure 8 shows the locations of the survey points and Table 2 shows the summary of the survey. Assuming the accuracy of handy GPS positioning is 50 m, that is to say equivalent to 3 pixels of ASTER VNIR spatial resolution, a bottom index value of the target point was calculated by mean value of 7 by 7 pixels surrounding the center pixel of the given coordinate as the target point. Inside of brackets of Bottom Index column in Table 2 represents number of non-null pixels in 7 by 7 windows. Calculated bottom index has been evaluated with the field data and summarized as follows:

- Bottom Index greater than or equal to 2.5 seems to correspond to sandy bottom, based on the comparison between substrata and Bottom Index.
- However, bottom type with Bottom Index greater than 2.4 may also be the sandy bottom (No. 1 & No. 6 in Figure 9), the detailed study may be needed to reveal the relationship between the coverage of the vegetation and the Bottom Index.

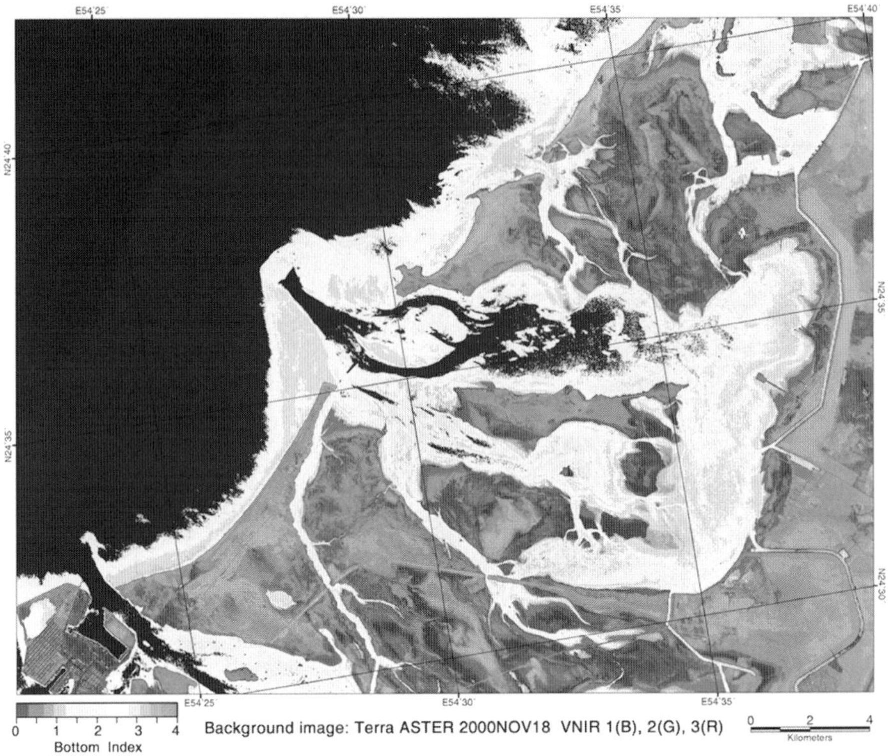

Figure 6. Example of bottom index image.

- The evaluation suggests that time difference of approximately 3 years between the field observation and satellite data acquisition made the accurate correlation between the field data and image analysis difficult.

4.2. Vegetation index to vegetation map

Discrimination of the vegetation areas from non-vegetation areas such as desert, salt pan, sabkha, bare soil, water, etc., is obviously important for environmental monitoring. Multi-temporal change of the vegetation distribution cannot be monitored without identification and extraction of vegetation areas. Vegetation indices are widely used for this purpose, as well as for rough estimation of vegetation activity and/or healthiness. In this study, threshold value discriminating vegetation from non-vegetation was sought by comparing the vegetation distribution map produced from Terra ASTER data with the Natural Resources Map (ERWDA, 2000). To find an appropriate threshold value, the mangrove areas were digitized from the Natural Resources Map first, and then SAVI values of those areas were examined. Salt marshes in the Natural Resources Map were not digitized because they likely include non-vegetated areas based on different criteria. By taking minimum value of SAVI for digitized areas as a tentative threshold value,

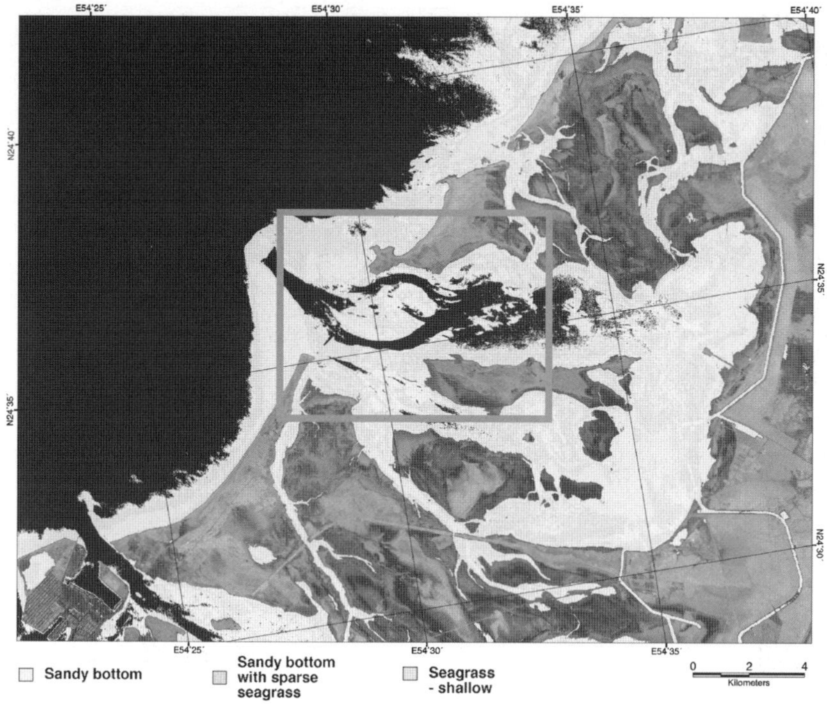

Figure 7. Provisional bottom type map.

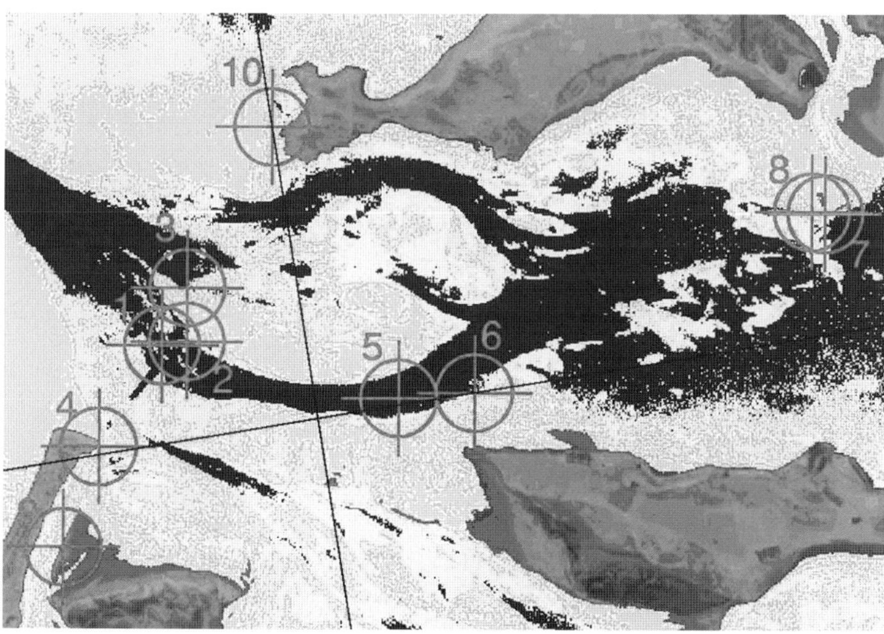

Figure 8. Locations of the field survey displayed on Terra ASTER provisional bottom type map. Locality shown in Figure 7.

Table 2. Results of the field survey.

No.	Longitude	Latitude	Habitat	Substrata	Water depth (ft)	Water depth (m)	Notes	BI*
1	54.4849	24.59254	Seaweed, shallow	Sandy	9	2.7432	Water turbid, seaweed sp. *Padina minor* (30% cover)	2.412(3)
2	54.48755	24.59254	Sandy bottom	Sandy	14.5	4.4196	Water turbid, high current, part of degraded channel?	2.422(21)
3	54.48848	24.59754	Sw/Seagrass shallow	Sandy	8.5	2.5908	*Padina minor* (10%) and *Halodule uninervis* (40%)	2.465(31)
4	54.47671	24.5841	Mixed sandy beach	Sandy	0	0	Sand, dead marine shells and gravel mixed beach	2.767(29)
5	54.50869	24.58402	Seagrass — Deep	Sandy	21.8	6.64464	*Halodule uninervis* (60%)	Null
6	54.51655	24.58335	Seagrass shallow	Sandy	5.4	1.64592	*Halophila ovalis* (10%) *Halodule uninervis* (30%)	2.442(39)
7	54.5558	24.59457	Seagrass shallow	Sandy	5.9	1.79832	*Halodule uninervis* (30%)	3.208(48)
8	54.5558	24.59457	Seagrass shallow	Sandy	0	0	Algal mats at periphery	2.566(49)
9	54.47143	24.57531	Tidal flats	Sandy	0	0	Close to Ras Al Gurab, sparse algal growth	2.686(13)
10	54.49993	24.61121	Rocky bottom	Rocks and Sand	7.5	2.286	Gravel, boulders and small sized rocks	2.636(49)

* 7 × 7 BI's are averaged. Bracketed number shows the number of non-null pixels in the 7 × 7 window.

Figure 9. Photographs of survey localities. (No photographs for localities No. 2, 3, 7 and 10.)

pixels having values larger than or equal to this threshold were extracted as vegetation pixels. Then the threshold was adjusted so that vegetation areas showing reddish color on the VNIR false color composite were completely extracted. Figure 10 shows an example of extraction of vegetation areas in the same region as Figure 5. It reveals that the areas of mangroves and salt marshes of the Natural Resources Map are satisfactory extracted. Mangrove areas seem to correspond to the areas with SAVI over 0.2 (Fig. 11). From theoretical point of view, however, it is impossible to discriminate the mangroves from salt marshes merely by the magnitude of SAVI. SAVI values bigger than 0.2 are likely resulted from high density of distribution of trees.

Although the extraction of vegetation areas is a mandatory process for monitoring of coastal environment, vegetation types must be classified from operational viewpoint because multi-temporal change in the distribution of certain vegetation type is focused in the course of actual monitoring. Vegetation types are relatively simple such as mangroves, halophytes other than mangroves, and algal mats in the coastal region of Abu Dhabi. Mangroves consist of single species *Avicennia marina* in the study area. *Arthrocnemum macrostachyum* is representative species of the halophytes except mangroves. Because each category has specific signature on the ASTER VNIR false color composite, discrimination among categories will be successfully made by the classification with appropriate

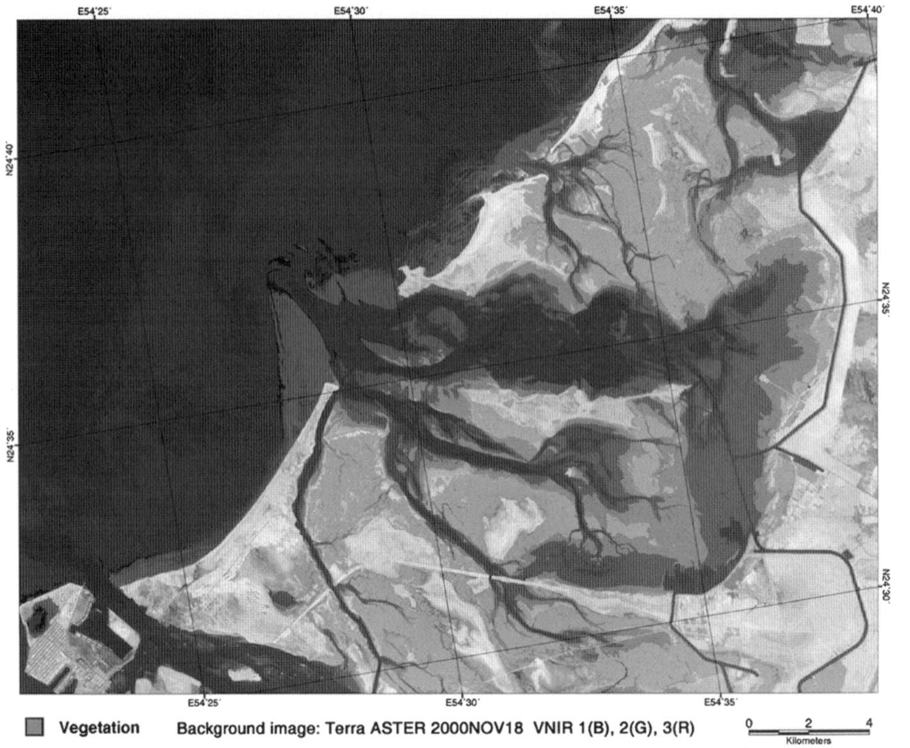

Figure 10. Example of extracted vegetation areas. Red pixels have SAVI values larger than or equal to 0.09.

supervisors. Areas of mangroves and salt marshes (halophytes) on the Natural Resources Map (ERWDA, 2000) were used as the supervisors (Fig. 11). Supervised classification using 9 bands of ASTER VNIR through SWIR was performed for the vegetation areas that were extracted by using certain threshold value of SAVI. Distribution of classified mangroves and halophytes seems to be fairly concordant with the Natural Resources Map (Fig. 12). However, field verification shows that, when trees are relatively low and sparsely distributed, they are not recognized as vegetation (Figs 12 and 13), suggesting one of limitations on the spatial resolution of 15 m. In turn, this may mean that dense forest of relatively taller mangroves is classified as mangroves by supervised classification.

5. Summary

Operational algorithms for producing bottom type and vegetation type maps have been developed using Terra ASTER image data in Abu Dhabi coastal region. Bottom type can be classified by using bottom index, which is calculated mainly from water-leaving radiance of Terra ASTER VNIR bands 1 and 2. An algorithm for extracting water-leaving

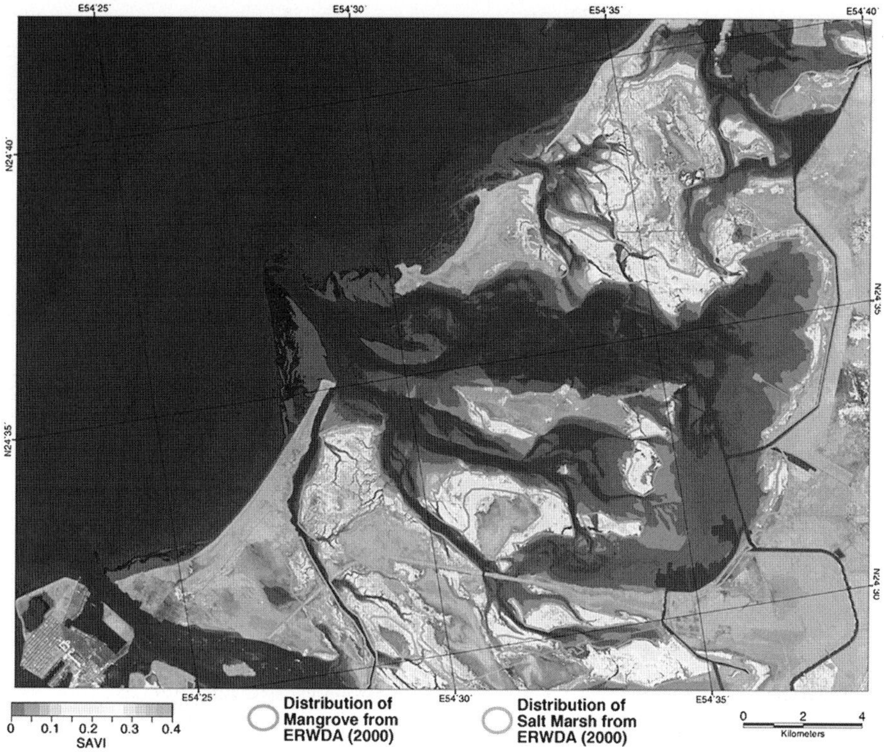

Figure 11. SAVI image map superimposed on Terra ASTER VNIR false color composite with mangroves and salt marshes from Natural Resources Map (ERWDA, 2000).

radiance accounting water surface reflection and sun glitter effect, together with corrections of atmospheric scattering, from signal received by ASTER has also been developed. Bottom index is calculated by

$$\mathrm{BI} = \ln(\mathrm{Lw}_1) - 0.44\ \ln(\mathrm{Lw}_2)$$

where Lw_1 and Lw_2 are water-leaving radiance for VNIR bands 1 and 2, respectively. By the comparison of bottom index image and results of sea truth performed by ERWDA, BI = 2.5 was provisionally deemed to be a threshold between sand and sea grass/seaweed. Sand has bottom indices more than or equal to 2.5 while sea grass/seaweed are less than 2.5.

The evaluation suggests that time difference of approximately 3 years between the field observation and satellite data acquisition made the accurate correlation between the field data and image analysis difficult. The result also suggests a field observation synchronized with satellite observation is necessary to obtain more accurate results. In addition, it became clear that a large number of field samples would be required to analyze accurately. In this pilot study some uncertainty is still remained because of limited number of the field samples.

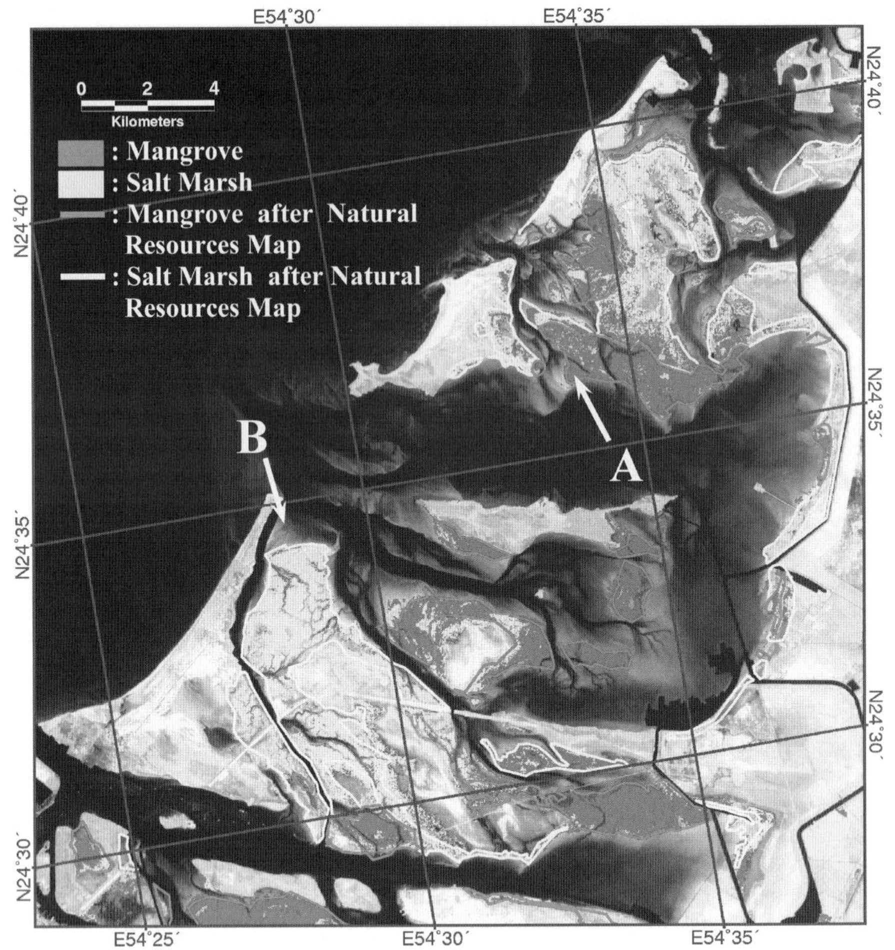

Figure 12. Classification result of mangroves and halophytes.

A (Height 2-6m)

B (Height < 2m)

Figure 13. Field photographs of mangroves. Locality shown in Figure 12.

Onshore vegetation is easily discriminated from bare soil, water, artificial construction, etc., based on SAVI. Simple atmospheric correction specially modified for Terra ASTER data was applied to convert DN values into the reflectance. After screening out of non-vegetation by SAVI, supervised classification using ASTER VNIR and SWIR bands has been performed for onshore vegetation to classify vegetation types such as mangroves and halophytes (except mangroves). As mentioned above, a future field observation is needed to be synchronized with satellite data acquisition for verifying the algorithm adopted in this research.

Acknowledgements

This study is a part of the collaboration research between Japan Oil Development Co., Ltd. (JODCO) and the Environmental Research and Wildlife Development Agency (ERWDA), called "ASTER Remote Sensing Alliance (ARSA)". Authors would like to express their thanks to ERWDA and JODCO for granting permission to submit this paper and to UAE. University for providing the opportunity to present the paper at OPEIC 2003.

References

Chavez, P.S. Jr., 1988. An improved dark-object subtraction technique for atmospheric scattering correction of multispectral data. Remote Sens. Environ. 24, 459–479.

Elterman, L., 1968. UV, Visible, and IR Attenuation for Altitudes to 50 km, 1968, Environmental Research Papers, no. 285, AFCRL-68-0153. Air Force Cambridge Research Laboratories, Cambridge, MA, p. 56.

Environmental Research and Wildlife Development Agency (ERWDA), 2000. Abu Dhabi Coastline Oil Spill Protection Priorities 2000, Natural Environmental Resources and their Oil Spill Protection Priorities for the Coastline of Abu Dhabi Emirate, U.A.E. ERWDA, U.A.E, p. 120.

Gordon, H.R., Clark, D.K., Brown, J.W., Brown, O.B., Evans, R.H., Broenkow, W.W., 1983. Phytoplankton pigment concentrations in the Middle Atlantic Bight: comparison of ship determinations and CZCS estimates. Appl. Opt. 1, 20–36.

Haltrin, V.I., 1998. Empirical Relationship Between Aerosol Scattering Phase Function and Optical Thickness of Atmosphere Above the Ocean. Propagation and Imaging Through The Atmosphere II. Proceedings of SPIE Vol. 3433, pp. 334–342.

Lyzenga, D.R., 1981. Remote sensing of bottom reflectance and water attenuation parameters in shallow water using aircraft and Landsat data. Int. J. Remot. Sens. 17, 155–166.

Ouaidrari, H., Vermote, E.F., 1999. Operational atmospheric correction of landsat TM data. Remot. Sens. Environ. 70, 4–15.

Takemura, T., Nakajima, T., 2002. Single-scattering albedo and radiative forcing of various aerosol spacies with a global three-dimensional model. J. Clim. 15 (4), 333–352.

Zhang, M., Carder, K., Muller-Karger, F.E., Lee, Z., Goldgof, D.B., 1999. Noise reduction and atmospheric correction for coastal applications of landsat thematic mapper imagery. Remot. Sens. Environ. 70, 167–180.

Oil Pollution and its Environmental Impact in the Arabian Gulf Region
M. Al-Azab, W. El-Shorbagy and S. Al-Ghais, editors

Chapter 3

Operational remote sensing for the detection and monitoring of oil pollution in the Arabian Gulf: case studies from the United Arab Emirates

Salem Essa[1,*], Hussein Harahsheh[2], Manabu Shiobara[3] and Takashi Nishidai[4]

[1]Geology Department, Faculty of Science, UAE University, P.O. Box 17551, Al-Ain, UAE
[2]Global Scan Technologies, P.O. Box 1286, Dubai, UAE
[3]Japan Oil Development Co., Ltd., 1-21-2, Shinkawa, Chuo-ku, Tokyo 104-0033, Japan
[4]JGI, Inc., 1-5-21, Otsuka, Bunkyo-ku, Tokyo 112-0012, Japan

Abstract

Oil spill pollution is a serious threat to marine environments along the UAE coast line. Considerable oil spills have been caused by accidental and deliberate oil sludge dumping from passing ships during the past. Damage to fishery, seawater desalination, and damage to plants and natural habitats are of real concern. The purpose of this study is to investigate and demonstrate the feasibility of using different types of satellite imagery for detecting oil spills in the Arabian Gulf, offshore the UAE. A number of satellite data were processed and analyzed by using images from early US intelligence satellites of the mid-seventies to the most recent earth observation satellites, European ENVISAT, 2003.

The analysis focused not only on oil spill but also on various appearances of oceanographic phenomena. A practical method for discriminating oil spills is an invaluable tool for marine oil spill surveillance. It is confirmed that the offshore of the United Arab Emirates faces frequent occurrences of oil spills in the Arabian Gulf and in the Gulf of Oman. Discharged oil and widely spread oil slicks in offshore Fujairah were confirmed by a sea truth campaign carried out in early 2003. An Oil Spill Atlas is being edited and is due to be published by the UAE University this year. The work done is the first step towards oil spill monitoring of the offshore UAE and its adjacent waters.

Near-real time monitoring using commercial satellites is thought to be feasible. Future earth observation satellites including the forthcoming Japanese ALOS and internationally operating earth observation satellites in conjunction with marine numerical modeling techniques are expected to be very efficient tools for marine pollution surveillance in the coming future.

Finally, an operational monitoring system integrating near-real time satellite observation with GIS mainframe is an ultimate goal that would constitute the nucleus for an Early Warning System to protect the marine environment of the Gulf States against oil pollution.

1. Introduction

Vast quantities of oil lie beneath the Arabian Gulf and the surrounding Gulf Coast Countries. The unique landscape of long parallel lines of folded ridges and valleys in the

*Corresponding author. *E-mail address:* salem.essa@uaeu.ac.ae (S. Essa).

Zagros Mountains of western Iran is created by the subduction of the Saudi Arabian plate by the Eurasian plate. Little vegetation can be seen in this part of the world (Fig. 1). The mountainous Musandam Peninsula that juts northward into the Strait of Hormuz, the entrance to the Arabian Gulf from Indian Ocean, is Rus al Jibal, Oman. The Strait of Hormuz has been a strategic focus in world affairs for thousands of years. In the last few decades, some 25% of the world's oil production has passed through the strait. This heavy use of the Gulf waters by oil tankers imposes real dangers facing its fragile environment.

Practically, the influx of oil from tankers and offshore oil operations are one of the major causes of pollution in the marine environment. Ballast water and other oily water discharged into the Arabian Gulf ranged from 400,000 to 750,000 ton in 1986. According to statistics of the US Coast Guard (1990), sources of oil in the sea are categorized into six categories (Fig. 2). By far the highest contributor to oil pollution in the ocean, about 52%, results from a mix of materials and wastes made up of urban runoff and land-based industrial discharges. Another 21% of the oil in the sea is directly attributable to the world's oil industry, where 2% of this occurs in spills from rigs and platforms during the exploration and production phases. Only 5% of oil pollution in the oceans is attributable to accidents involving oil tankers. However, one big spill may seriously damage the life in the sea and coastal areas in the short term. The remaining 13 and 9% of hydrocarbons in the oceans are absorbed by the atmosphere by particle settlement and rain-wash or attributable to natural seepage, respectively.

Figure 1. Vast quantities of oil lie beneath the Arabian Gulf and the surrounding Gulf Coast countries captured in this synoptic northwest-looking (from International Space Station, acquired on 3 March 2003).

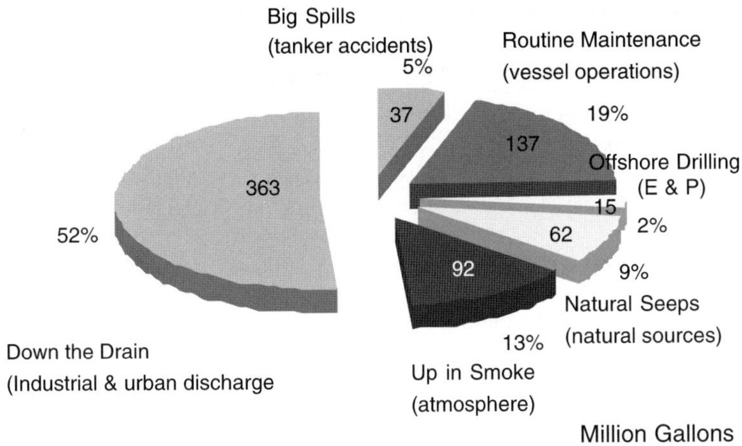

Figure 2. Sources of oil in the sea (after US Coast Guard, 1990).

The environment in the Arabian Gulf region is also under considerable threat from intentional or accidental oil spills, ballast water discharged, dredging and infilling for coastal development, and uncontrolled sewage and industrial wastewater discharges. Oil discharged from ships imposes a much greater long-term threat to the marine environment than one big accident. Monitoring illegal oil discharges is thus an important component in ensuring compliance with marine protection legislations and general protection of the coastal environments. The study areas were selected as evaluation sites for demonstrating oil spill appearance in daily operations and for testing the resolution necessary for oil slick characterization.

We focused on discriminating between direct hydrocarbon pollution such as big spills by tanker accidents, discharged oil caused by routine maintenance, and leaking oil from offshore exploration and development operations by means of satellite imagery. Satellite-borne sensors have different electromagnetic characteristics with varying limitations for detecting marine surface features; therefore, a combination of sensors is required to monitor marine oil pollution effectively. This case study indicates that most oil spills are found along the major shipping routes and in anchorage area as well as in such areas with intensive large-scale oil production activities with leakage or tank-washing discharges. The results will help in locating potentially vulnerable areas, and serve as a reference in future routine monitoring. There have already been several remarkable accidents involving the loss of large quantities of crude oil from disabled tankers (Table 1). Even a small spill can cause havoc in the ecologically sensitive environment.

During the last 10 years, satellite-borne synthetic aperture radar (SAR) sensors have been broadly used for oil spill monitoring and have provided excellent data. Their advantages encompass the capability of wide coverage, under any weather conditions all day and night; the near-real time data delivery; and the improved cost-efficiency when compared with airborne SAR surveillance. Some very successful examples of ERS SAR applications in marine oil spill detection have been reported, such as the near-real time monitoring in Norway (Pederson et al., 1995), monitoring oil spill pollution with ERS

Table 1. Major oil spill incidents offshore UAE.

Date	Volume of spilled oil (tons)	Oil type	Location
2001/01/14	1300–1500	Fuel	1 mile offshore Jabal Ali
2001/01/24	300–900	Heavy	7 mile NE offshore Abu Dhabi
1998/01/07	5000–10,000	Crude	5 mile offshore Ajman
1994/03/30	16,000	Light	9.6 mile offshore Fujairah

SAR in the Mediterranean (Pavlakis et al., 1996) and the conviction in an oil-spill case in Singapore (Lu et al., 2000).

2. Reliability of SAR systems

The development and use of an airborne/spaceborne remote sensing system for surveillance of the sea surface, detecting the pollution and monitoring for their spreading from space is required. Such system must provide all-weather observation, be independent of illumination conditions and cloud cover, define the position, type and volume of oil spill, and work in real scale of time (Witte, 1986). SAR is an active remote sensing tool in which a satellite antenna transmits microwave signals towards the ocean surface. The interaction between the sea surface and microwaves is very sensitive to variations in sea surface roughness. Rough surfaces scatter large amount of energy back to the antenna and have bright signatures while smooth surfaces reflect the energy away from the antenna and have dark signatures. Since short surface waves (ripples and capillary waves) are usually present on the water surface, it effectively scatters microwaves via the Bragg scattering mechanism (Valenzuela, 1978). It is well known, that crude oil and other oil substances form films of various thicknesses on the sea surface. Oil films locally damp sea surface roughness and give dark signatures (dark patches among brighter surrounding surface), so-called slicks, on the SAR images (Huhnerfuss et al., 1981; Alpers and Huhnerfuss, 1988). This fact gives a physical basis for application of spaceborne radars for oil spill detection and monitoring in the ocean (Huhnerfuss et al., 1981).

The behavior of oil spill on the sea surface significantly depends on its important physical–chemical properties, such as viscosity, density, surface tension, and elasticity. Moreover, crude oil is a complex mixture of different chemical components including heavy and light fractions. Typically, crude oil during its evolution in the sea can be detected in different phases, which are listed here in order of their age: oil spill, oil film, emulsion, blue shine, and aggregates (Kotova et al., 1998). During the lifetime of oil spill in the sea it will be exposed to a number of processes, which dramatically influence physical–chemical properties. Termed weathering, these processes are as follows: spreading, drift, evaporation, dispersion, emulsification, bacterial degradation, and photo-oxidation (Kotova et al., 1998). With time the physical–chemical properties of oil spills are changed due to the effect of these processes. These processes play an important role in oil spill detection by using spaceborne SAR.

The detectability of oil spills in SAR images strongly depends on the wind speed at the sea surface. Under low wind speeds, typically between 0 and 2–3 m/s, the sea surface looks dark on SAR images. In this case the wind-generated waves are not already developed and oil films look dark on a dark background; detection in this case is impossible. Wind speed between 3 and 6 m/s is ideal for oil slick detection, the sea surface roughness is developed and oil slicks appear as dark patches on a bright background. However, when wind speed reaches 10–12 m/s, detectability is impossible again or obstructed due to the redistribution of oil spills/slicks by the surface waves and wind-induced mixing in the upper ocean layer (Scott, 1986). As a result, slick disappears from the sea surface and SAR imagery.

Other detection problem is discrimination between man-made and natural organic film slicks (Huhnerfuss et al., 1986). Natural biogenic films of a very thin thickness resulting from life-cycle of plankton and other marine organisms can form slicks on the sea surface and, in turn, produce similar dark signatures on the SAR images (see Gade et al., 1998, and references herein).

3. Study areas and oil pollution

Study areas lie between longitudes 52°E and 56°45′E, latitudes 24°15′N and 26°N, it covers two sub-areas, one offshore of Abu Dhabi to the northeast up to Ras al Khaimah Emirate in the Arabian Gulf and the other offshore Fujairah in the Gulf of Oman (Fig. 3). The Arabian Gulf is about 990 km long with maximum width of about 338 km. It is a shallow sea-oriented in NW–SE direction with an opening connected to the Gulf of Oman through the Strait of Hormuz. The average water depth is about 36 m. Evaporation and wind are major driving forces of water circulation in the Arabian Gulf. Evaporation is stronger in winter due to high wind speed, than summer when the water surface temperature is higher.

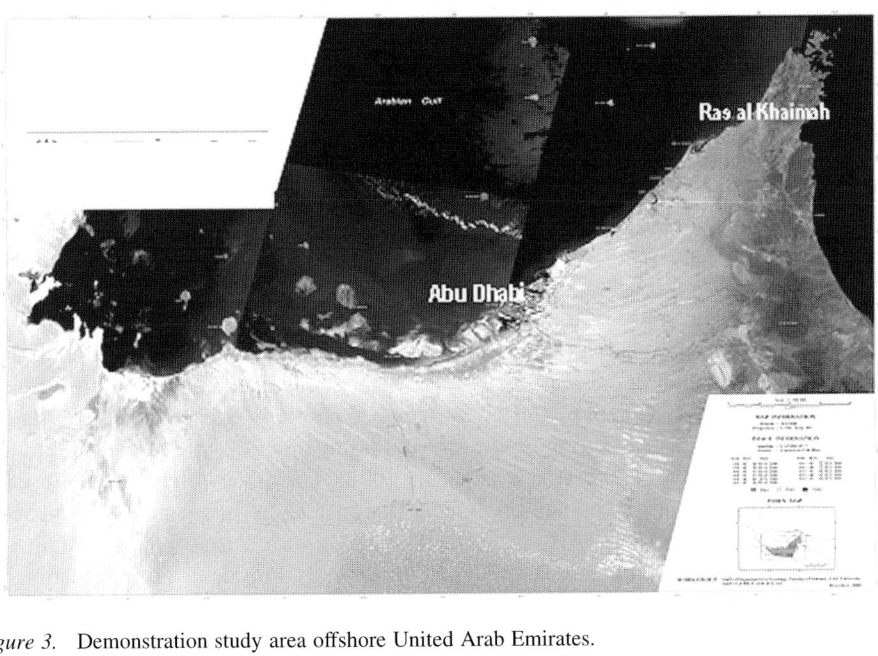

Figure 3. Demonstration study area offshore United Arab Emirates.

The overall circulation in the inner Arabian Gulf is cyclonic, with relatively fresh water entering through the Strait of Hormuz. In winter, the cyclonic circulation is primarily found in the southern portion. This cyclonic circulation gradually expands its influence farther north during summer time. Tides are mostly diurnal and semi-diurnal and are not significant for the residual circulation, except near topographic features. The tides are complex standing waves with two amphidromic points for the semidiurnal constituents (northwest and southeast) and one for the diurnal constituents (central, near Bahrain). Three current patterns are used to simulate the circulation in the Arabian Gulf (Chao et al., 1992). Theses include a reverse estuarine flow, river flow, and a surface water circulation driven by a NNW wind.

The Arabian Gulf region is the largest offshore oil development area in the world. Bahrain, Iran, Iraq, Kuwait, Qatar, Saudi Arabia, and the United Arab Emirates (UAE) produced over 27% of the world's oil in 2000. The area also holds 65% of the world's oil reserves. The study area have one of the busiest and most important tanker shipping lanes in the world; one ship passes the strait approximately every 6 min, another statistic indicates that more than 40% of the world's total oil transportation passes through the region. About 15.5 million barrels of oil per day is transported through the Strait of Hormuz. Contamination influx is mainly from tankers releasing ballast, tank cleaning leakage from drilling rigs and production platforms, and ship accidents.

4. Satellite image processing project

For strengthening ties of friendship between Japan and UAE, the Japan Oil Development Co., Ltd. (JODCO) with the collaboration of Japanese Information Center for Petroleum Exploration and Production (ICEP) have supported the implementation of the: Satellite Image Processing Project (SIPP). SIPP aims to contribute to the development of the basic infrastructure, management and operation of satellite image processing system in the United Arab Emirates University – Remote Sensing Lab., and to assist the Remote Sensing Lab crew with the development of the practical use of remote sensing data for environmental protection. Petroleum industry in the UAE, in particular, has been challenged by environmental issues in recent years, and has been required to advise appropriate countermeasures in the event of offshore oil spill.

In this atmosphere, the UAE University and Japanese group proceeded research collaboration between April 2000 and March 2004 on application of satellite image data on environmental monitoring and oil spill detection.

SIPP achieved the study of oil spill hazards on the west and east coasts of UAE and presented its results to several international conferences. We note the contribution of SIPP to the international conference on "Integrated Management of Marine Environment in Arabian Gulf" 21–23 October 2001, with a paper entitled "Sea Surface Observation Using Satellite Imagery for Marine Environmental Protection Offshore of the United Arab Emirates: Preliminary Results" (Alsharhan et al., 2002). The final work has been presented during the international conference on "Oil Pollution and Its Environmental Impact in the Arabian Gulf Region" Al-Ain, 5–7 October 2003, with a paper entitled "Oil Spill Detection Using Satellite Imagery in Offshore United Arab Emirates". The production of a national atlas of Oil Spills of UAE coastal areas will commemorate the final output of this project and is due to be published before the end of 2004.

Table 2. Satellite imagery investigated during the study project.

Sensor	Number
JERS-1 SAR (Japan)	64 scenes
Shuttle Imaging Radar C/X-SAR (US)	37 segments
ERS-1/2 AMI (EU)	15 scenes
RADARSAT SAR (Canada)	3 scenes
ENVISAT ASAR (EU)	7 scenes
Hexagon KH-9 (US)	12 scenes
Landsat-7 ETM + (US)	15 scenes
JERS-1 OPS (Japan)	92 scenes
Terra ASTER (Japan/US)	95 scenes
Shuttle Handheld Camera Photographs	50 photos

5. Data set

More than 300 satellite images have been examined during this study (Table 2). ERS-1/2, RADARSAT, and ENVISAT C-band SAR data has been used for the great majority of oil spill detection. However, other satellite images even optical sensor images have shown good detection capabilities. Therefore, we looked for all available image data archives and selected more than one hundred of images derived from different platforms that covered the most of the offshore waters of the UAE. We conducted a search of ERS-1 and 2 data archive to compile a list of all images acquired over the study areas. To evaluate their suitability for slick detection, historical wind conditions for corresponding SAR images were obtained. For each acquisition date, surface wind speed histories were reconstructed using historical records. Because radar backscatters from the sea surface are strongly affected by surface wind patterns, the wind speed histories were used to rate the suitability of each image for oil slick detection.

A base map for the entire UAE, onshore and offshore, was made using 11 Landsat-7 ETM + images (Table 3, Fig. 4). Some field pictures of historical oil contamination in the UAE waters were obtained from NOAA historical oil spills information. Oil well location data were also used to discriminate ships and oil production facilities. The images were processed to emphasize slick appearance and provide accurate location information, and were incorporated into a database containing well locations. The appearance of a small oil slick emanating from a production platform strongly supports the concept of using radar imagery to operationally detect small leaks, but a test demonstration over a well-recognized site may be required to assess this technology.

6. Image analysis and oil slick detection

Each satellite-borne imaging sensor behaves differently in oil slick detection. Briefly speaking, in the visible and near-infrared region, the absorption and reflection of solar illumination determines the characteristics of the oil by means of spectral reflectance of electromagnetic energy. Crude oils show different colors, i.e., spectral reflectance varies

Table 3. 11 Landsat 7 ETM + were used to create the Mosaic for the onshore and offshore UAE.

Satellite	Sensor	Date
Landsat-7	ETM +	19/May/2000
Landsat-7	ETM +	26/May/2000
Landsat-7	ETM +	28/May/2000
Landsat-7	ETM +	02/June/2000
Landsat-7	ETM +	04/June/2000
Landsat-7	ETM +	09/June/2000
Landsat-7	ETM +	27/June/2000

depending upon chemical composition of crude oil. Sea surface roughness changes the direction of solar illumination reflected from water, due to randomly scattered sun glint from wave facets oriented at the specular angle, as does the presence of sea surface condition shown in Figure 5 (Berry, 1995). As a result, reflectance contrast between clear

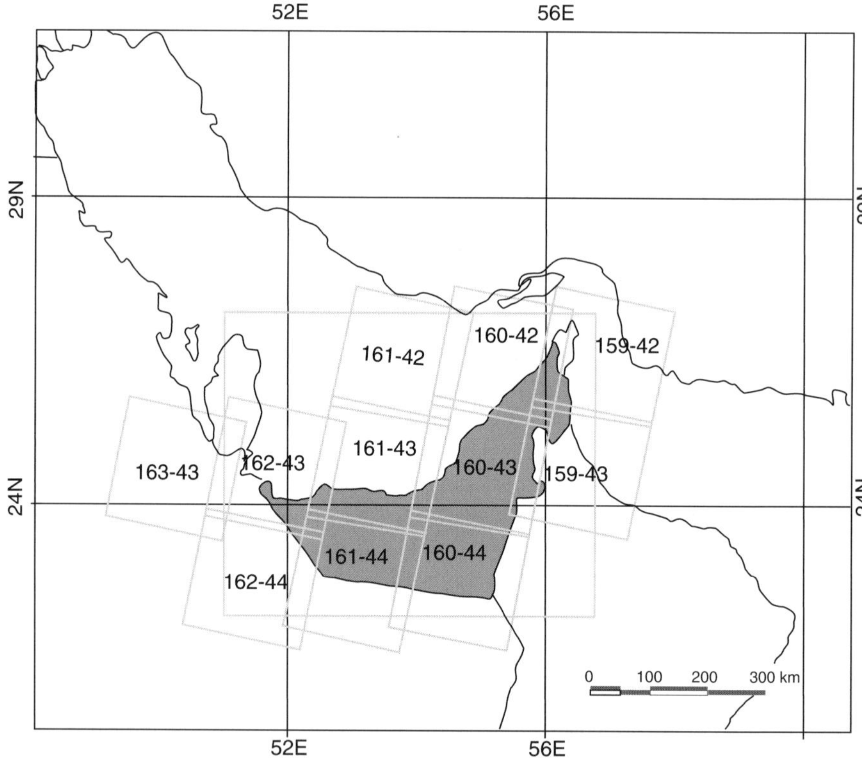

Figure 4. Index map of the Landsat-7 ETM + mosaic image of the entire United Arab Emirates onshore and offshore area shown in Figure 3.

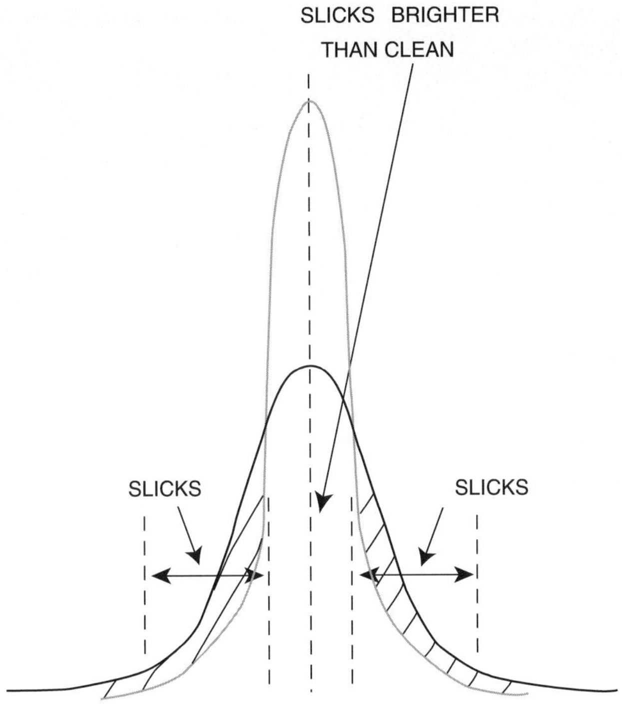

Figure 5. Appearances of oil slicks on the sea surface. Slicks are brighter than clean water for Glint Angle (the specular angle between the direction of the reflected ray and the direction of the satellite) less than 10°. Slicks are dark for Glint Angle between 10° and 40°, and not resolvable for angles more than 40°. This is only true for current velocity less than 2 m/s and wind velocity less than 10 m/s.

water and oily water varies with sea state at any given wavelength in the visible and near-infrared region (Fig. 6).

By looking at the thermal infrared region, the water surface temperature can be calculated. Brightness temperature is calculated with respect to the function of both emissivity and kinetic temperature. A film of oil on water has the same kinetic temperature as the water since they are in direct contact.

The difference of emissivity 0.02 between seawater and crude oil makes an apparent difference of 1.2°C in brightness temperature in the wavelength region of Landsat TM band 6 at room temperature (Salisbury et al., 1993). Generally, an oil-slicked surface shows a lower temperature than the surrounding clear water surface (Fig. 7). The image data observed at night time is more reliable to avoid the influence of solar illumination difference between seawater and oil slicks.

In the microwave region, see Section 2 above, the brightness of the sea surface is a measure of backscatter of the sea surface roughness. As smooth sea surface appears dark and the brightness increases as the sea surface becomes rougher. Oil films damp wind-generated gravity capillary waves on the sea surface. Hence they appear dark against brighter surrounding areas in a SAR image.

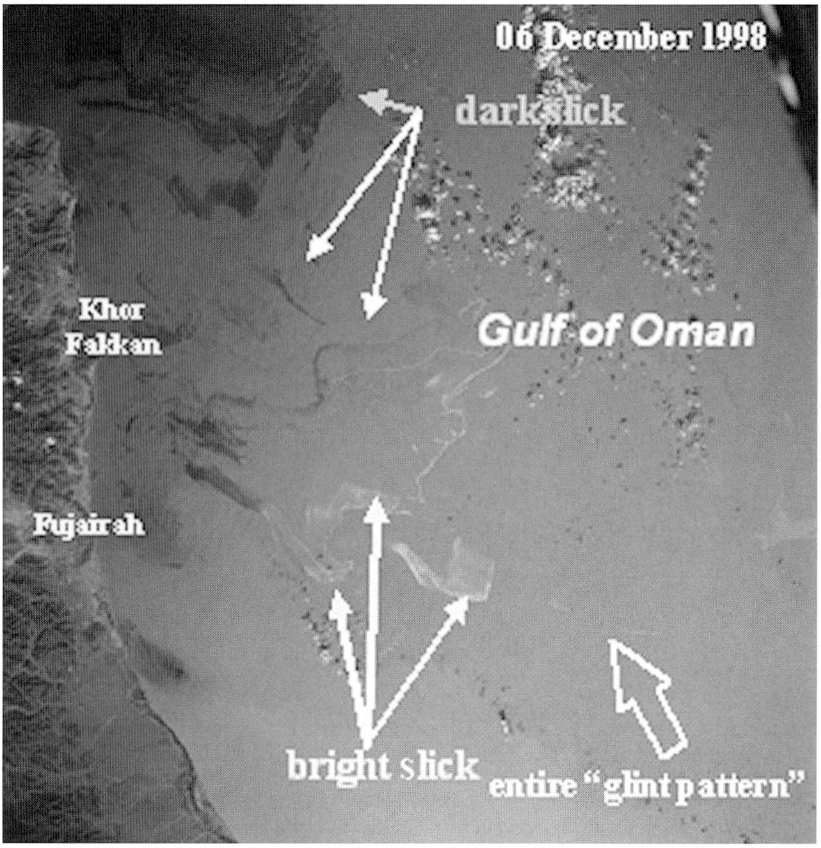

Figure 6. Shuttle photograph offshore Fujairah in the Gulf of Oman, showing the entire glint pattern, with slicks bright against a relatively dark background in the center (blue arrow), and dark against an even darker background around the edges (white arrow).

Sea state information such as a two-dimensional wave height and direction (WAM of Naval Oceanographic Office), and weather information such as wind history information (hindcast) is important for oil spill monitoring using both optical and SAR satellite imagery. The reliability of an analysis result is highly influenced by availability of these data. Some of the marine surface features are easily recognized with the assistance of general weather information or due to their distinctive shape and configuration. Generally speaking a hindcast is the most important to discriminate slicks by oil spill from natural seepage or film slicks, wind shadow, and algal bloom (Espedal and Wahl, 1999).

A large archive of historical satellite radar imagery exists for the Arabian Gulf region stretching back to 1984. However, not necessarily all of these images are useful for showing oil slicks. Indeed, the suitability of radar imagery for slick detection is largely a function of the local weather and sea-state conditions at the time of image acquisition. The image data used in this study were geographically transformed to fit the base map image

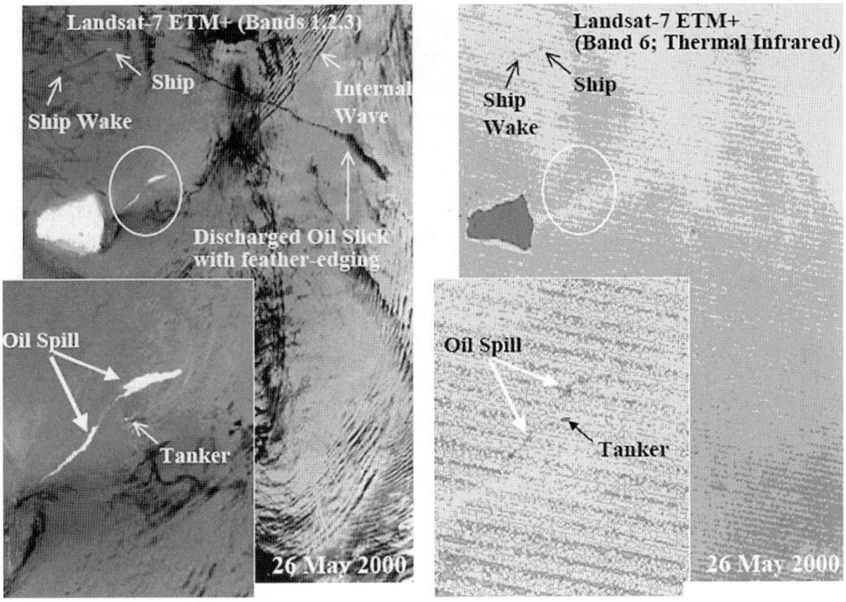

Figure 7. Comparison between Landsat-7 ETM + visible bands composite and thermal infrared band images.

using corner point locations provided with the images. Images brightness and contrast were manipulated to optimize the discrimination of slicks.

After the radiometric range correction and the georeferencing of each image, the procedure starts with the definition of a target area. For each ERS SAR image, we obtained hindcast wind speed data at every 6 h interval spanning the 24-h prior to the time of satellite's overpass. Wind speed at the time of image acquisition was between 3 and 5 m/s. Ideal conditions were considered to be: a wind speed of 3–4 m/s at acquisition time, with a maximum speed well below 10 m/s and average speed well below 5 m/s during the prior 24 h. As a minimum, several hours of fair conditions are required for new slick to accumulate to a detectable size. If at any time during the previous 24 h the wind speed exceeded 10 m/s, then a thin oil slick will most likely have been dispersed. If the wind speed exceeded 5 m/s during the previous 24 h, it will most likely have prevented the formation of a thin oil slick. The ERS-1/2 images used in this study were observed during suitable wind condition for slick analysis.

Analytical efforts were focused on overall assessments of image quality and suitability for slick detection, identification of possible oil slicks in the area, and comparison of slick patterns between the different dates of imagery and also different types of sensors. The interpretation results indicate that certain coastal areas of the UAE face frequent oil spills. Figure 8 shows striking examples, one offshore Fujairah (centered at the coordinates 25°30′N/56°25′E). Here considerable spill concentrations were found within successive JERS-1 OPS, Landsat-7 ETM + images and ERS-1/2 SAR browse images. Figure 8 is a time sequence images comparison of the 29 June 1992, 21 May 1995, and 28 May 2000

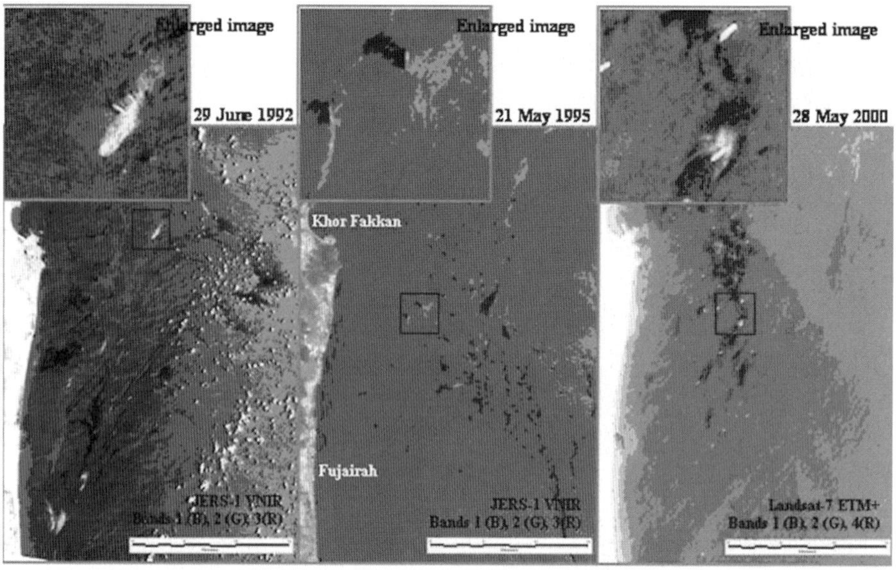

Figure 8. JERS-1 OPS and Landsat-7 ETM + time sequence images, offshore Fujairah, oil tankers anchorage points in the Gulf of Oman.

images. Oil discharged from both anchored and moving vessels can be observed in each image. Immediately after discharging flush ballast water, the simmering water surface can be seen as bright silver to gray color patches on the surrounding water. Based on the size of the image pixel, most of the vessels are super tankers whose size of hull is more than 300 m in length.

Fujairah Port Authority has introduced Fujairah Offshore Anchorage Area (FOAA) since February 1993 to restrict and prohibit anchoring in the area from Bidiya (north of Khor Fakkan) to Dibba. A Sea-truth campaign was jointly carried out in the area of FOAA and surrounding areas by the teams of the United Arab Emirates University and Japan Oil Development Company with cooperation of the UAE Coast Guard. Ship-borne observation and sampling sea surface and seafloor materials were carried out. Chemical analyses of the sample materials were conducted.

Interpretation of these results is under investigation to reach a conclusion on the extent of oil pollution in the UAE offshore area, especially the FOAA, and its impact on the local environment. An example of tanker accident oil spills, shows how long oil contamination has remained in the beach even oil clean-up scheme were applied immediately after the incident (Fig. 9). The accident happened on 30 March 1994, when two super tankers, UAE tanker BAYUNA and the Panamian-registered SEKI, collided 17 km off the coast of Fujairah, several thousand gallons of crude oil washed ashore on beaches between Khor Fakkan and Dibba in the Emirate of Fujairah. Roughly 16,000 tons of crude oil spilled into the sea, spreading over a total of 40 nautical square miles. Photos taken in February 2003 show remnants of weathered oil 9 years later.

Shipping routes with relatively frequent incidences of oil spills are the offshore area running parallel to the coast of Abu Dhabi, Dubai, Sharjah, and Ajman where intensive oil

Figure 9. Oil contamination remained in the beach near Khor Fakkan even oil clean-up scheme were applied immediately after the incident.

production activities exist, and the routes through the Strait of Hormuz. This study is probably the first time that a spatial distribution of oil pollution is mapped over quite a wide sea area in the region using high spatial resolution satellite images. Attention was focused on monitoring natural seepage of oil in the area with coordinates 25°15′N/54°00′E of the offshore of Abu Dhabi. The analysis was carried using images from, ERS-2 SAR, and RADARSAT-2 SAR images. Figure 10 shows numerous small oil slicks caused by natural seeps from the seafloor throughout the images.

A partial problem is the effect of extensive dark, low wind areas across the northeastern parts of the 29 and 30 May images. This weather pattern is not uncommon in the Arabian Gulf region. Scattered oil spills are marked by dark patches with a variety of sizes and shapes on the sea surface. Despite these localized problems, both images were acquired under generally acceptable conditions for slick detection. It is seen clearly that there are

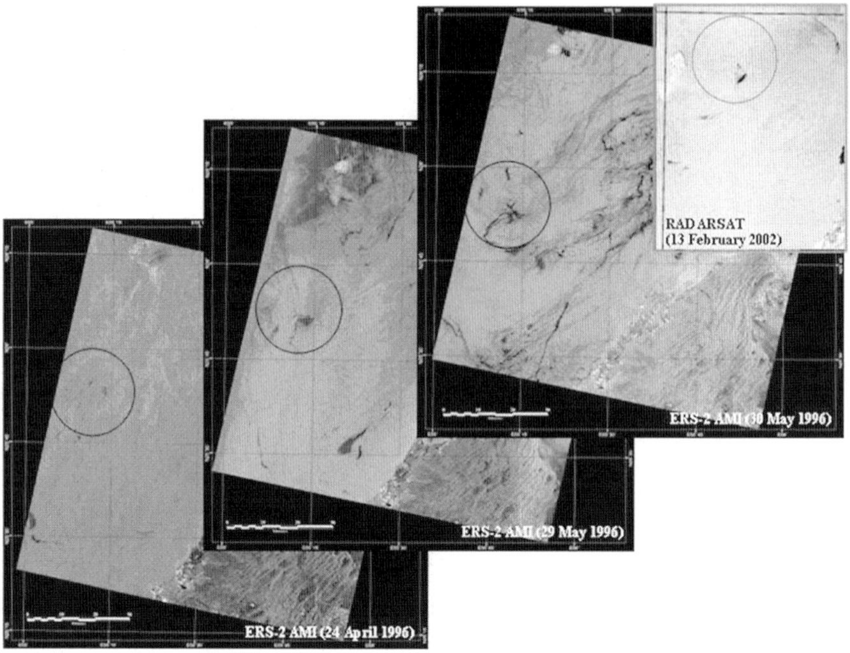

Figure 10. Multitemporal SAR (ERS and ENIVISAT) image of leaking oil from oil production facilities (circle) offshore UAE.

two polluted regions, one in the west side of the image and one in the southwest corner of the image. If we put this map over the layer representing the oil fields and the shipping routes, a close relationship between spills and both oil fields and the shipping routes can be found visually.

Adjacent to the oil field offshore Ras Al Khaimah, near the left edge of the image (Fig. 10, circle), many distinct slicks appear with different shapes, which derived from leaking oil from oil production facilities. Observed slicks in the three images acquired on different dates are confirmed as leakage oil slicks from the same oil production platforms. In addition, locations of existing wells correspond to locations of the leaking points. The clear discrimination of these slicks on these images provides strong evidence that imaging conditions were well within bounds for reliable oil slick detection. One additional feature of interest appears on the 29 May image (Fig. 11); the platform is obviously trailing a small, wind-driven slick that extends 10 km to the southwest and the bend nearly perpendicular to ESE. This bend seems to be due to a change of surface current. Oil platforms and vessels appear as bright spots, generally indistinguishable unless the well locations are superimposed on the images. A few vessels can be distinguished by the presence of short, faint lines that indicate a trailing ship wake. The size of the longest slick detected was 25 km. The characteristics suggest that the slicks are ephemeral surface events from a moving source, probably oil spills from a passing vessel moving down from north to south.

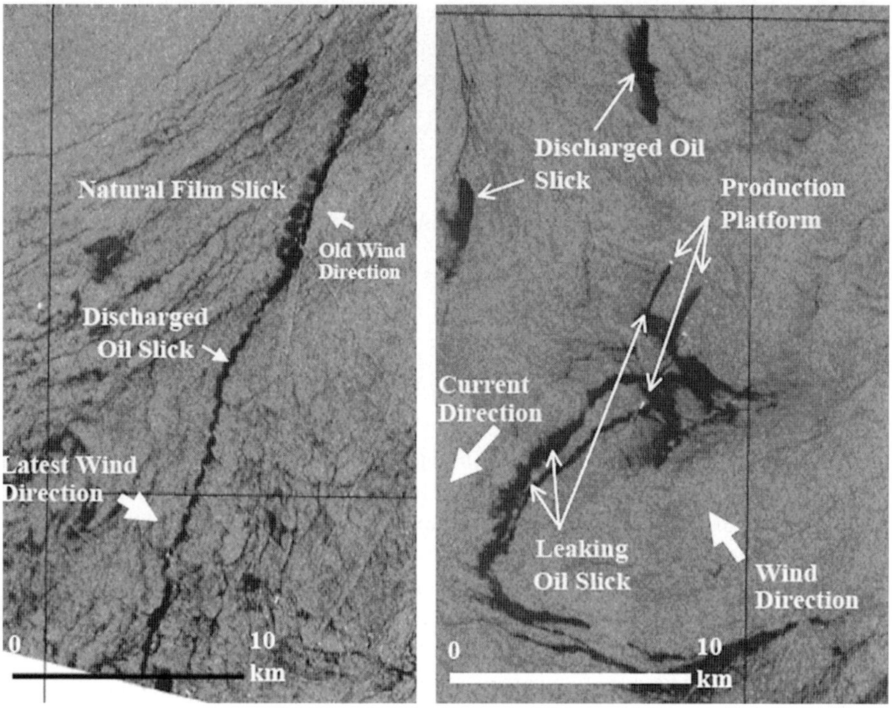

Figure 11. Typical slicks patterns of discharged oil from moving ship and leaking oil from production facilities. Image acquired on 29 May 1996.

7. Conclusions

Results of this study demonstrate and confirm that the offshore UAE faces frequent occurrences of oil spills both in the Arabian Gulf and the Gulf of Oman. In particular the offshore Fujairah, the Gulf of Oman has considerable spill concentrations found in multi-temporal image analysis. It is thought to be caused by high oil content ballast water discharged from giant oil tankers. It is worth stressing here that the identification of such areas is an important step for planning intense monitoring scheme based on spaceborne imagery, which is attainable with the current acquisition conditions. The results of this study have confirmed that satellite imagery can detect oil spill under certain conditions. Detection capabilities of ERS, RADARSAT and ENVISAT SAR satellites are good for thin pollutants (< 1 μm) and wind speeds at 3–4 m/s, they still can detect thick emulsions at wind speeds up to 18 m/s. Sources of other detectable pollutants detected (besides petroleum) are:

- Run off water from onshore acid depository
- Drilling fluid/produced water from oil rigs
- Waste from fish production plants
- Fish oil remaining on the sea surface after trawler catches and processes
- Palm oil

The biggest advantage of using satellite-borne sensors is that large areas can be covered in one shot and the accumulated information from the satellite data analysis, then can guide airborne surveillance efficiently. However, detection limitations are:

- Natural slicks are frequently observed at low wind speeds. Experienced interpreters or advanced pattern recognition algorithms are needed to discriminate between oil slicks and look-alike artifacts
- At high wind speeds the pollution will be mixed into the water column: no dampening effect expected

Satellite limitations

- Verification of pollution type and volume is not possible
- Current limitation by satellite coverage and price policy – most customers have small environmental protection budgets

The environment sensitivity database initiated by Abu Dhabi National Oil Company (ADNOC) (Jensen et al., 1992) has continued to be compiled and revised by Environmental Research and Wildlife Development Agency (ERWDA) (Perry et al., 2001). Creating an integrated coastal zone pollution monitoring system including GIS that combined satellite information with this database should be one of the top priority concerns. Considerable work will undoubtedly be needed before all the possibilities of satellite imagery are fully realized for oil spill monitoring across the worlds maritime routes. However, there have been many encouraging signs. The Gulf countries, such as the UAE, Kuwait, and Qatar, have initiated efforts to use satellite imagery. Satellite radar imagery will be essential for future oil spill detection. Planned future satellites, RADARSAT-2 and Japanese ALOS in conjunction with other widely available satellite data will become decisive data sources for oil spill monitoring, early alert and prevention systems.

The essential information that any monitoring should have is: oil spills/false alarms database, wind and weather information, map and database with all the potential static sources of pollution (oil platforms, industries, etc.), main ship routes, sea currents and seepage locations. Furthermore, to have an appropriate oil spill emergency program, some effort should be put into getting a regional, unified system for ordering different satellite data, fast data and product distribution and special data policies.

Operational usage has however not been fully developed except in some special cases, such as the one established in Norwegian offshore and coastal regions. The daily activity of pollution monitoring services with the integration of satellite remote sensing technology in some European countries are important steps, which need to be made in widening their applications and building initiatives with other maritime regions including the Arabian Gulf and the Gulf of Oman. About a hundred of ERS SAR, RADARSAT, Landsat TM, and Terra ASTER data were acquired and archived during the last 5 years over the study area and surrounding regions. They constitute a substantial resource for this purpose. Every sensor has its advantages and disadvantages, and therefore it is necessary to combine sensors for satisfactory results. An integrated GIS system using different data types and linking all the Gulf countries will constitute the basis for an Early Warning System to protect these countries' offshore and coastal regions of fragile environments against oil pollution in the long run.

Future development and improvements are:

- Better satellite instrumentation and algorithm
  - ALOS (Japan)
  - Earth Explorers (ESA)
  - RADARSAT II (Canada)
  - TerraSAR X, L (Germany, UK)
- High spatial resolution satellite (spatial resolution of 1–4 m)
  - IKONOS (USA)
  - Quick Bird (USA)
  - EROS-1 (USA)
  - Orbiview-3 (USA)
- Integration in GIS: Future developments should focus on delivering real-time information on coastal marine environments via dynamic GIS-based system by using advanced fusion techniques for variable data sources (*in situ*, satellite data) with integrated marine simulation algorithms.
- Modeling: Focusing on making water dynamics models more reliable for coastal waters.

Acknowledgements

This study is a part of the collaboration project between Japan Oil Development Co., Ltd. (JODCO) and the United Arab Emirates University, called "Satellite Image Processing Project (SIPP)" under the support of the Information Center for Petroleum Exploration and Production, Japan. Authors would like to express their thanks to all those who participated in the accomplishment of this study both in Japan and the UAE.

References

Alpers, W., Huhnerfuss, H., 1988. Radar signatures of oil films floating on the sea surface and the Marangoni effect. J. Geophys. Res. 93, 3642–3648.

Alsharhan, A., Shiobara, M., Yamaguchi, M., Onuma, T., Nishidai, T., 2002. Sea surface observation using satellite imagery for marine environmental protection offshore United Arab Emirates. Proceedings of the Seventh International Conference on Remote Sensing for Marine and Coastal Environments, 20–22 May 2002, Maiami, Florida, 9 p.

Berry, J.L., 1995. Detecting and evaluating oil slicks on the sea surface. The Proceedings of the Space Congress on Remote Sensing for Oil Exploration and Environment, Bremen, Germany, 23–24 May, pp. 90–110.

Chao, S.Y., Kao, T.W., AL-Hajri, K.R., 1992. A numerical investigation of circulation in the Arabian Gulf. J. Geophys. Res. 97 (C7), 11219–11236.

Espedal, H.A., Wahl, T., 1999. Satellite SAR oil spill detection using wind history information. Int. J. Remote Sens. 20(1), 49–65.

Gade, M., Alpers, W., Huhnerfuss, H., 1998. Imaging of biogenic and anthropogenic ocean surface films by the multifrequency/multipolarization SIR-C/X-SAR. J. Geophys. Res. 103, 18851–18866.

Huhnerfuss, H., Alpers, W., Fast, O., 1986. On the discrimination between crude oil spills and monomolecular sea slicks by airborne remote sensors – today's possibilities and limitations. Proceedings of IGARSS'86, Zurich, 8–11 September, pp. 1359–1364.

Huhnerfuss, H., Alpers, W., Jones, W.L., 1981. The damping of ocean surface waves by a monomolecular film measured by the wave staffs and microwave radars. J. Geophys. Res. 86, 429–438.

Jensen, J.R., Narumalani, S., Weatherbee, O., Michel, J., Hayes, M.O., Murday, M., Sexton, W.J., Davis, B.A., Green, C.J., 1992. Utilization of remote sensing and GIS technologies for oil spill planning and response: case studies in Florida, United Arab Emirates, and Saudi Arabia. The Proceedings of the First Thematic Conference on Remote Sensing for Marine and Coastal Environments, New Orleans, Louisiana, June 15–17, pp. 145–151.

Kotova, L., Espedal, H.A., Johannessen, O.M., 1998. Oil spill detection using spaceborne SAR: a brief review. Proceedings of the 27th International Symposium on Remote Sensing Environmental, June 8–12, Tromso, Norway, pp. 791–794.

Lu, J., Kwoh, L.K., Lim, H., Bao, M., 2000. Mapping oil pollution from space. Backscatter 11 (1), 23–26.

Naval Oceanographic Office, Model WAM, Significant wave height and significant wave direction 48 hours forecast.(http://www.navo.navy.mil/cgibin/graphic.pl/metoc/59/21/0-0-1/0).

Pavlakis, P., Sieber, A., Alexandry, S., 1996. Monitoring oil-spill pollution in the Mediterranean with ERS SAR. Earth Obs. Q. 52, 8–11.

Pederson, J.P., Seljelv, L.G., Wahi, T., Skoelv, A., 1995. The Norwegian oil spill detection project – from R&D preoperational early warning detection service. The Proceedings of the Space Congress on Remote Sensing for Oil Exploration and Environment, Bremen, Germany, May 23–24, pp. 126–138.

Perry, R.J.O., Drew, C.R., Gliddon, D.J., 2001. Natural environmental resources and their oil spill protection priorities for the coastline of the Abu Dhabi Emirate, UAE. The Abstracts of the International Conference on Integrated management of Marine Environment in the Arabian Gulf, October 21–23, Al-Ain, UAE, pp. 36–37.

Scott, J.C., 1986. Surface Films in Oceanography. ONRL Workshop Report, C-11-86, Office of Naval Research, London, pp. 19–34.

Salisbury, J.W., D'Aria, D.M., Sabins, F.F. Jr., 1993. Thermal infrared remote sensing of crude oil slicks. Remote Sens. Environ. 45, 225–231.

US Coast Guard, 1990. Update of inputs of petroleum hydrocarbons into the oceans due to marine transportation activities, National Research Council. National Academy Press, Washington, DC.

Valenzuela, G.R., 1978. Theories for the interaction of electromagnetic and oceanic waves – a review. Bound. Layer Meteorol. 13, 61–85.

Witte, F., 1986. Oil slick detection with a side looking airborne radar. Proceedings of IGARSS'86, Zurich, September 8–11, pp. 1369–1374.

Oil Pollution and its Environmental Impact in the Arabian Gulf Region
M. Al-Azab, W. El-Shorbagy and S. Al-Ghais, editors

Chapter 4

Remote sensing applications for coastal and marine resources management

Hussein Harahsheh

Global Scan Technologies, P.O. Box 1286, Dubai, UAE

Abstract

The capability of satellite remote sensing to provide synoptic, repetitive and multispectral data has proved to be very useful in inventory and monitoring of coastal features, such as tidal wet-lands, potential aquaculture sites, mangroves, estuary-dynamics/shoreline-changes, and off-shore aspects like suspended sediment dynamics and coastal currents, near-shore bathymetry, internal waves, etc. The Indian Remote Sensing Satellite data IRS 1C 1D and IRS-P4 (Oceansat) have been found to be useful in study of the above coastal features. It also has potential towards marine applications by providing chlorophyll/primary productivity estimate and extent and spread of oil slicks. Significant results of these applications and studies are outlined in this chapter. Thematic maps of coral reefs are being prepared using IRS 1C LISS data for the Gulf of Kach. Change detection results are being produced using IRS 1D to show the changes on the Earth's surface before and after Gujarat earthquake, while IRS-P4 OCM data are used to detect the pre- and post-distribution of chlorophyll concentrations and suspended solids concentrations.

1. Introduction

Marine environment occupies a very vast part of the Earth's surface and the economy of mankind has been deeply associated with the ocean. They are always important, but they are even more important now, because population is increasing (60% of the world population live there, i.e. 3 billion people) and land resources become limited. Hence the marine environment attracts the individuals, institutes and government agencies all over the world, who are concerned about the state of art in this region. This region attracts large number of research scientists also who are concerned about the problems and solutions towards a peaceful and sustainable use of the ocean. So besides conventional methods used for scientific studies which lack wide and repetitive data collection in a short span of time, the modern and latest tool like remote sensing, GIS, and GPS are used for their unique capabilities of providing repetitive and synoptic coverage over the vast coastal regions; they are also relatively cheaper, used in cloudy and rainy days and

E-mail address: husseinh@belhasa.ae (H. Harahsheh).

provide real time picture of natural hazards occurring all over the world. Hence, remote sensing (RS) application is the need of the hour for sustainable development and management of the marine environment.

This chapter is a compilation of various remote sensing applications and other integrated information systems in the study of various ocean parameters, features, processes and in the exploration and exploitation of various marine resources all over the world.

2. The technology of remote sensing

2.1. Definition

Remote sensing is a technology to observe objects' size, shape and character without direct contact with them. Also remote sensing is the science dealing with the acquisition, processing and interpretation of images and related data obtained from aircrafts and satellites that record the interaction between matter and electromagnetic radiation. When electromagnetic radiation falls upon a surface, some of its energy is absorbed, some is transmitted through the surface, and the rest is reflected. Surfaces also naturally emit radiation, mostly in the form of heat. These reflected or radiated electromagnetic waves are received by sensors aboard Earth observation satellites. Since the intensity and wavelengths of this radiation are a function of the surface in question, each surface is describes as processing a characteristic "spectral signature." If an instrument can identify and distinguish between different spectral signatures, then it will be possible to map the extent of surfaces using remote sensing. In general, the characteristics of reflected or radiated electromagnetic waves depend on the type or condition of the objects. Therefore, by understanding the characteristics of electromagnetic waves and by comparing to the observed information, we can know the size, shape and character of the objects.

2.2. Characteristics of remote sensing

There are some characteristics of remote sensing by earth observation satellites. Applying these characteristics, the observation data are utilised for various studies related to the earth environment as well as various fields affecting to our lives. Characteristics of remote sensing by earth observation satellites are:

2.2.1. Enables to observe broad area at a time

The next image shows Amman/Jordan and surrounding area of about $140 \times 140\ km^2$ taken by the Indian IRS 1D satellite on 12th May 2003 and received by Global Scan Technologies receiving station in Dubai. The swept area differs among sensors aboard satellites. The observation by satellites can cover a broad area at one time, and it is very useful to find out land use, vegetation distribution, ruins or structures' shape and size.

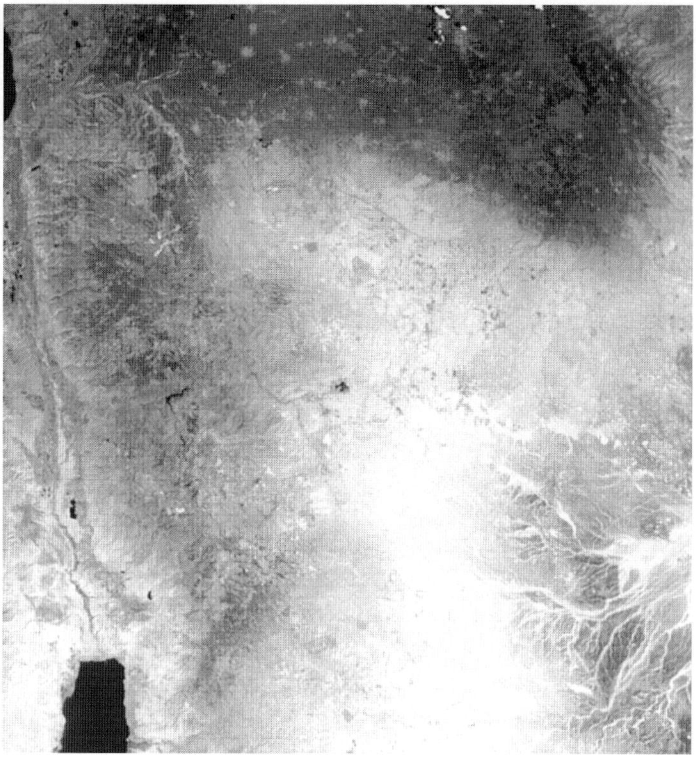

2.2.2. Long period observation

The following images are describing the real time development of the great palm island of Dubai between February and June 2003; this figure shows the big extension of the palm island, observed by Indian remote sensing satellite IRS 1D panchromatic mode. Earth observation satellites orbit the Earth repeatedly, so they can observe the same area regularly. Therefore, we can see the secular changes of environment.

2.2.3. Enables us to know the condition without visiting the area

Earth observation satellites observe various areas, so it is useful to find out environmental changes and damage status of the areas where we cannot easily go or when a natural disaster occurs. These IRS 1D LISS-III images show an area around Abrantes at the river Tejo in central Portugal. The image on the left was acquired on 29th July 2002. The acquisition on the right is from 8th August 2003 when forest fires were still burning. The area shown in the sample is approximately $70 \times 100\ km^2$ (Source: www.euromap.de).

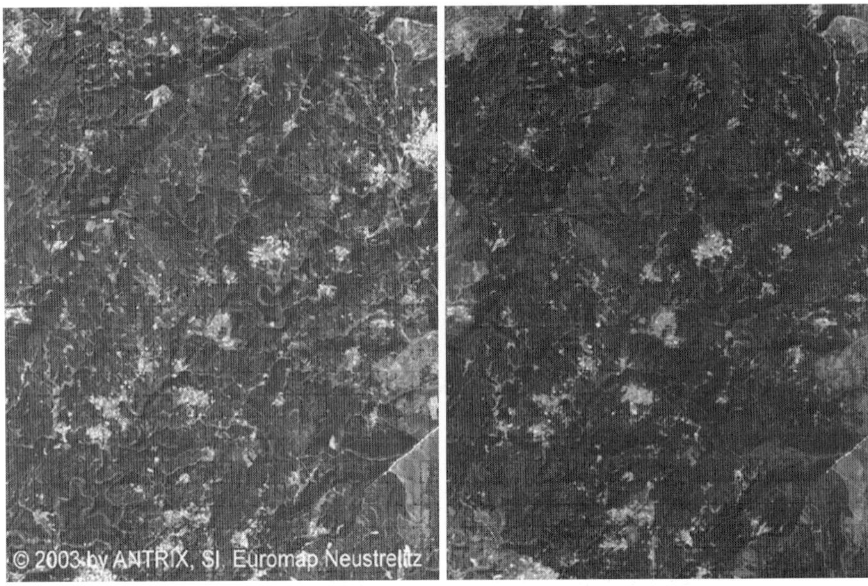

2.2.4. *Invisible information*

Observation by satellites can acquire invisible information like temperature, and has provided new discoveries to the Earth environmental study. This figure shows sea surface temperature map calculated using MODIS image.

2.2.5. *Lower cost*

When compared to other data collection methods including aerial photography, land surveying, etc. Scientists can work with quite old images which are considerably cheaper.

2.2.6. *Spectral information*

The spectral range of data available cannot be match and provides a wealth of information.

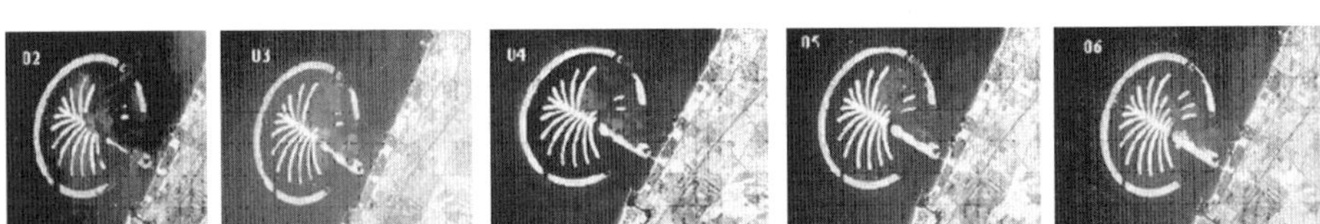

2.2.7. *Holistic dataset*

A holistic dataset is provided in one image showing the interaction of several natural phenomena.

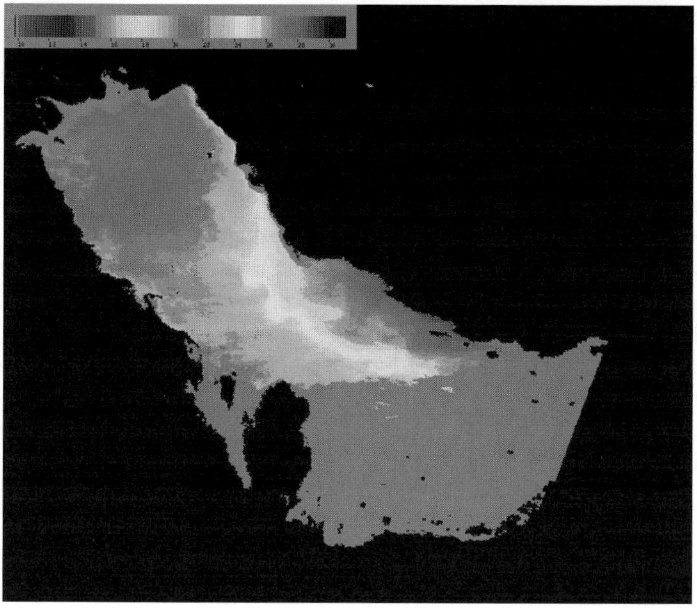

2.3. Applications of remote sensing

Remote sensing applications have proved very useful in resource surveys and management activities. These techniques have been applied successfully in forestry, agriculture, disaster management, flood and drought monitoring, land use and land cover mapping, urban planning, mineral targeting, environmental impact assessment, coastal zone mapping, etc. Application of remote sensing techniques to coastal and marine research has been now extended for the identification of potential fishing zone (PFZ), estimation of primary productivity using ocean colour data (Green, 1996), identification of geomorphological changes of beaches, mapping coastal wetlands, distinguishing coral seagrass and mangrove habitats and monitoring marine pollution. The launch of IRS satellites series is yet another milestone in the space science history of the world. Different sensors of these satellites offer unique application opportunities and would near time information system for resources including marine wealth.

3. Ocean colour monitor (OCM) on Indian Remote Sensing Satellite IRS-P4

While the earlier IRS series of satellites carried cameras (LISS-I, II, III, WiFS & PAN) designed predominantly to meet the needs of land-based applications, IRS-P4 is the first Indian satellite envisaged to meet the data requirements of the oceanographic community. The payload flown on-board IRS-P4 are: (a) ocean colour monitor (OCM) operating in eight narrow spectral bands in the visible/near-infrared region of the electromagnetic spectrum and with high revisit time (2 days) and (b) multi-frequency scanning microwave radiometer (MSMR) operating in microwave bands 6.6, 10.65, 18 and

21 GHz in dual polarisation mode. The MSMR is envisaged to provide information on physical oceanographic parameters such as sea surface temperature, wind speed and atmospheric water vapour. The IRS-P4 spacecraft will be a polar orbiting satellite in sun synchronous orbit with nominal altitude of 720 km, providing revisit time of 2 days for OCM.

The OCM instrument designed for the IRS-P4 satellite programme is significantly different from similar instruments built and launched to date. The main features of the OCM instrument are outlined in Table 1.

The OCM is the first instrument to take advantage of pushbroom technology for achieving higher radiometric performance and higher spatial resolution while maintaining a large swath to provide high revisit time for ocean observations.

The pushbroom approach has enabled the use of a 12 bit digitiser to cover the instruments dynamic range. The anticipated radiometric performance is shown in Table 2. The instantaneous geometric field of view of the pixel is 360 m across track and the sampling interval along track is 250 m. The instrument is mounted on a mechanism to provide tilt in the along track direction to avoid sun glint.

4. IRS data-coastal and marine applications

The IRS data in particularly IRS-P4 OCM (Oceansat) data would be extremely useful for the following marine and coastal features:

1. Estimation of phytoplankton in oceanic/coastal waters, production of global marine phytoplankton
2. Detection and monitoring of phytoplankton blooms
3. Chlorophyll/primary productivity
4. Coastal upwelling
5. Suspended sediment dynamics
6. Location of fronts, coastal currents, internal waves
7. Identification of water mass boundaries and
8. Pollution in the oceans
9. Spread of oil slicks
10. Salinity prediction
11. Sea surface temperature (SST)
12. Fishing and marine transportation
13. Oil and gas exploration and offshore industries
14. Global data on marine optical properties
15. Mangrove wetland monitoring and management
16. Coral reef studies such as mapping, identification, etc.
17. Near-shore bathymetry profiling
18. Coastal land use and their dynamic changes
19. Tidal wet-lands monitoring
20. Estuary-dynamics/shoreline-changes

With additional input from other sensors as well as conventional data, IRS-P4 OCM data will provide detailed information on the coastal region owing to its increased spatial

Table 1. Major specifications and features of IRS-P4: OCM.

Parameter		Specification
IGFOV at nominal altitude (m)		360 × 250
Swath (km)		>1420
No. of spectral bands		8
Spectral range (nm)		402–885
Spectral band	Central wavelength (bandwidth) in nm	
C1	414 (20)	35.5
C2	442 (20)	28.5
C3	489 (20)	22.8
C4	512 (20)	25.7
C5	557 (20)	22.4
C6	670 (20)	18.1
C7	768 (40)	9.0
C8	867 (40)	17.2
Quantisation bits		12
Camera MTF (at Nyquist frequency)		>0.2
Data rate (Mbits s^{-1})		20.8
Along track steering		+20, 0, 20

resolution. The information on pigments, in conjunction with sea surface temperature, will greatly assist in identification of potential fishery zones in coastal and oceanic waters. The potential end users of the OCM data products include fisheries management, marine industries, environmental management and studies related to the estimation of primary productivity in the oceanic basins. IRS-P4 OCM, along with other ocean colour sensors such as IRS-P3 MOS, SeaWiFS, MERIS and MODIS is assisting the ocean colour community in filling data gaps, and can also be used for the inter-calibration of different ocean colour sensors.

Table 2. IRS-P4 OCM — NESR/$\Delta\rho$ values.

	NESR (mw cm^{-2} sr^{-1} μm^{-1})	NE $\Delta\rho$ (%)
C1	0.026	0.047
C2	0.022	0.037
C3	0.017	0.026
C4	0.017	0.029
C5	0.015	0.026
C6	0.01	0.021
C7	0.005	0.014
C8	0.008	0.025

5. Remote sensing in mangrove research

Mangrove forests are one of the most important coastal ecosystems in the world in terms of primary production and coastal protection. Distributed in the tropical and sub tropical regions, mangroves reach their maximum development and greatest luxuriance in south-east Asia. Mangrove forests are now under stress in almost all tropical countries because of natural and demographic pressures. Along the Indian coasts, mangrove has been affected severely due to human induced stresses such as deforestation and other developmental activities. Therefore, there is a pressing need to have integrated approaches for coastal zone (mangrove) management as a means of achieving sustainable resources development. To study and evolve remedial measures to the extent possible, various organisations have been conducting a variety of research and planning, implementation and monitoring activities. Remote sensing technology is an important tool in this assessment because of its ability to provide synoptic view of the earth which would not be possible from the ground without exhaustive field surveys.

Wetland mapping should be done for the better understanding of various conditions of the wetlands and for the delineation of areal extent and boundaries of wetland especially coastal wetlands. These maps will serve as baseline data for classifying the coastal zones into preservation, conservation and development zones.

IRS LISS II Landsat TM data have also proved extremely useful for wetland mapping as well as for delineation high and low water lines. Likewise, it is possible to distinguish mangroves from other plant communities (Nayak et al., 1986; Nayak, 1993). Further, multi date satellite data could be used effectively to find out the changes in the extent of mangroves. For example, the 1986 TM and 1993 IRS LISS II data have helped to quantify the changes in the area cover of mangroves since both the sensors have similar resolution (Krishnamoorthy, 1997). False colour composites (FCC) derived from the green, red and infrared bands of satellite data can be virtually analysed on 1:25000 or 1:50000 scale when information is required at the state/national level. An image interpretation key indicating the tone colour, size, shape, texture pattern, location and association can be prepared for each category of vegetation including mangroves using ground truth information, topographical maps, aerial photographs, etc. The classification accuracy can be tested on a sample basis assuming binominal distribution for the probability of success/failure of sample tests. Sample size is decided using the look up table (LUT), prepared by employing a binominal probability model (Arnoff, 1982).

6. Case studies

6.1. Ocean changes as a result of natural hazards — Gujarat earthquake — study conducted by Ramesh O. Singh, Anjeeb Bhoi, Chandresh, Alok K. Sahoo and Rahul Kanwar from Indian Space Research Organisation (ISRO), India

6.1.1. Introduction

The Gujarat earthquake of 26th January 2001 generated intense shaking which was felt in 70% region of India. The intense shaking brought out changes in the ocean and land

Figure 1. View of the Earth's surface near Rapar after Gujarat earthquake.

surface and subsurface. At numerous places, emergence of water on the Earth's surface brought hope among people about the future availability of drinking water in Gujarat. The preliminary laboratory analysis of water samples collected in the epicentral area show high chloride concentration and salt content which ruled out any hope of the potable water. Figure 1 shows the view of the Earth's surface near Rapar which shows crater like structure which is common in volcanic area. The craters are seen on the surface from where the water has come out as a fountain up to about 5 ft high which was seen by local people and the water flow continued for about 2 h. Near to these craters, long cracks up to about 50 cm wide running about 500 m long with orientation east–west and northeast and southwest were seen. Nearer these craters, pebbles (crystals of silica) also emerged in water fountain due to pressure. Such observations conclude that this earthquake was associated with significant component of vertical force.

Multi sensors (IRS-P4 OCM, MSMR and IRS 1D) and multi date remote sensing data were acquired from National Remote Sensing Agency (NRSA), Hyderabad. Using the multi sensors and multi date data, various ocean and land changes have been mapped.

6.1.2. IRS-P4 Oceansat data

The IRS-P4 Oceansat is the first in the series of operational ocean remote sensing satellites which was launched on 26th May 1999. The OCM sensor records in eight bands at visible and near-infrared wavelengths. The field of view of the optics is 430, providing a swath of 1420 km from 720 km altitude. The digital data for two dates, one prior (18th January 2001) and after (26th January 2001 soon after) earthquake were analysed and chlorophyll concentration and suspended solid were deduced. The comparison of pre (Fig. 2a) and post (Fig. 2b) images shows drastic changes in chlorophyll concentrations and suspended solids in adjoining Arabian ocean especially in creek region of Gujarat and also along the western coast (Navalgund and Kiran Kumar, 1997).

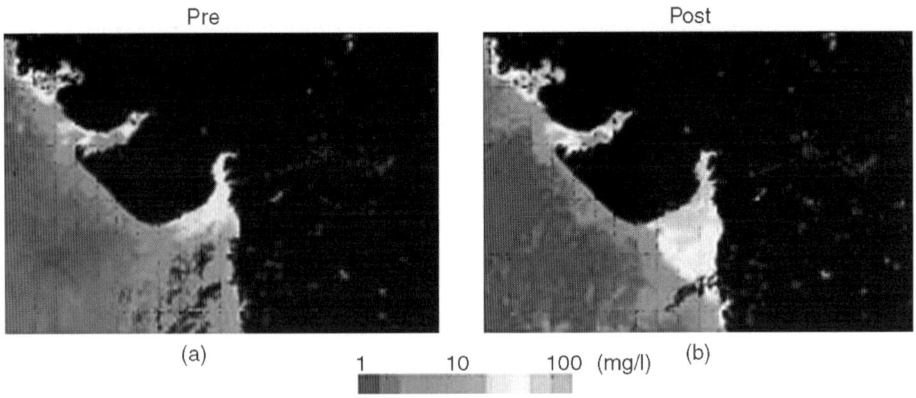

Figure 2. Pre- and post-distribution of chlorophyll concentrations using IRS-P4 OCM data.

The routine measurements of IRS-P4 OCM data may give some indication of the building of the stresses inside the Earth due to upwelling of the ocean. Such routine analysis is only possible if such data are provided to the potential users continuously and freely.

6.1.3. IRS-P4 MSMR data

IRS-P4 MSMR data were analysed prior and after the earthquake. Significant changes in brightness temperature over land and ocean region were noticed at two frequencies 6.6 and 10 GHz in both the vertical and horizontal polarisations. At some places, increase and at some places decrease in brightness temperature has been found. The significant changes in brightness temperature are likely due to the changes in moisture level. The changes in brightness temperature are also supported from the LISS III and OCM data which also show significant changes in the moisture content.

6.1.4. IRS 1D LISS-III data

IRS 1D LISS-III pre- and post-data were purchased from NRSA, Hyderabad. Surface features were extracted from these images and tectonic lineaments were extracted. The post earthquake image of 29th January shows significant surface manifestations in the epicentral area. The comparison of Figure 3a,b reveal that the moisture level has increased significantly due to which the paleo channel is clearly observed from the post-image (Fig. 4). The surface manifestations were also observed at the same places in the field.

6.2. Change detection analysis of eastern coastal zone of India using Satellite Remote Sensing Sensor Data by Amit K. Bhattacharya and P. K. Srivastava, Indian Institute of Technology Kharagpur, India

The present chapter is intended to monitor changes in wetland habitats and adjacent uplands in eastern coastal zone of India at intervals of 5–10 year cycles. This type of

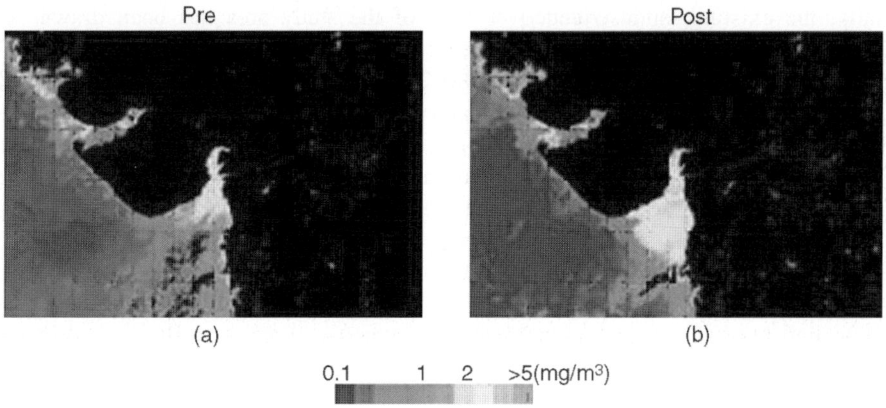

Figure 3. Pre- and post-distribution of suspended solids concentrations using IRS-P4 OCM data.

information and frequency of detection are required to improve scientific understanding of the linkages of coastal and submerged wetland habitats with adjacent uplands and with the distribution, abundance and health of living marine resources. Satellite imagery (primarily IRS-P3 WiFS and IRS 1B LISS II), aerial photography, and field data are interpreted, classified, analysed and integrated with other digital data in a geographic information system (GIS). The resulting land cover change databases are disseminated in digital form which can be used for conducting geographic analysis in the study area in future.

The shoreline within this area is approximately 500 km long and lies between latitude 18° 00′ to 21° 30′ N and longitude 84° 45′ to 87° 30′ E. Initially IRS-P3 WiFS data (path/row — 101/62) has been digitally processed using various image processing techniques, to monitor coastal land cover changes of the study area. However, detail landform changes have been carried out using high resolution IRS 1B LISS II satellite sensor data along with aerial photographs corroborated with selective field studies.

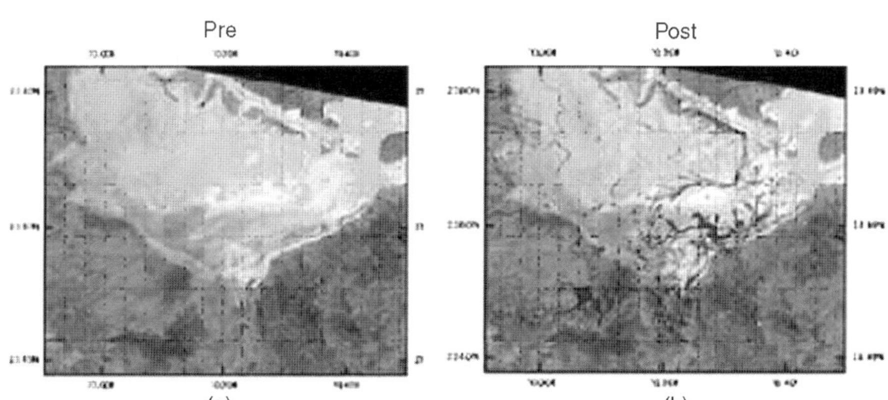

Figure 4. Pre and post distribution of suspended solids concentrations using IRS-P4 OCM data.

Finally, the existing landuse/landcover map of the study area has been drawn. This analysed result is compared with historic base maps acquired from Survey of India (SOI) topographic sheets of the period 1970–1980. The coastal land cover has been subdivided in to three super classes namely, (1) upland landcover, (2) wetland and (3) water and submerged land. These super classes are subdivided into class which is further subdivided into subclasses. The classification system focuses on landcover classes that can be discriminated primarily from satellite remote sensor data. Wetland super class consists of six classes, namely, marine/estuarine rocky shore, marine/estuarine unconsolidated shore (beach, flat, bar), estuarine emergent wetland, estuarine woody wetland, riverine unconsolidated shore (beach, flat, bar) and lacustrine unconsolidated shore (beach, flat, bar). Water and submerged land consists of those wetlands with surface water but lacking trees and shrubs. This study shows that IRS-P3 WiFS and IRS 1B LISS II data can be successfully used in studying the coastal landuse/landcover and landform of the eastern coast of India.

6.3. Case studies in mangrove using remote sensing by L. Kannan, T.T. Ajith Kumar, A. Duraisamy CAS in Marine Biology, Annamalai University, India

Estimates of mangrove cover are subjected to several sources of inaccuracy and confusion. If the mangrove canopy is dense, it obscures the treeless patches and channels so that the total mangrove cover is over estimated if the mangroves occur in either small patches or in low density. This is because small mangrove patches may be beyond the sensor's spatial resolution.

State	Area in sq km (Govt. of India Status Report)	Area in sq km (IRS data)
West Bengal	4200	1619
Andaman & Nicobar	1190	770
Orissa	150	187
Andhra Pradesh	200	480
Tamil Nadu	150	90
Gujarat	260	1166
Maharashtra	330	138
Goa	200	5
Karnataka	60	19
Kerala	Sparse	Sparse
Total	6740	4474

The total area of mangroves in India was estimated to be 6740 sq km (Status Report, Government of India, 1987). But the Indian Remote Sensing data have shown that the total mangrove area in India is 4474 sq km (Nayak, 1993) as detailed below:

The NRSA, Hyderabad has recorded a decline of 70,000 ha of mangroves in India within a period of 6 years from 1975 to 1981 and Vedaranyam/Point Calimere coastal areas of Tamil Nadu have lost 40% of their mangroves with a reduction of 18% of fishery resources within a period of 13 years from 1976 to 1989.

Assessment of the degradation of the Pitchavaram mangroves in Tamil Nadu has been possible to a large extent due to the information obtained from satellites. In 1987, the Pitchavaram mangrove forests were declared as reserve forests with an area of about 700 ha. Of this, nearly 62.8% of the mangrove area has been degraded between 1897 and 1994 as revealed by satellite data. Further, the remote sensing analysis has been much affected.

A comparison of the SOI topo sheet and satellite imagery of IRS 1B-LISS II shows that the breadth of the beach area around Pitchavaram has been reduced by 550 m between 1930 and 1970 and further about 150 m between 1970 and 1992 and the rate of erosion has been calculated as 13 m a year (Ajit Kumar et al., 1995). The comparison of topo sheet and satellite imagery also shows that erosion and sedimentation occurs simultaneously in the Pitchavaram area, which is a serious problem. If this trend is not changed, mangroves may be completely wiped out from here soon. So, a long term management plan is the immediate necessity to save these mangroves. The information required for mangrove management will be mainly on the distribution and extent of mangrove areas, forest composition, degradation sites, drainage network, spread of coastal villages, other land uses with in and outside the mangroves, etc. and it is essential to use high resolution remote sensing data to prepare large scale thematic maps on vegetation cover. For this, both remotely sensed data and GIS can be used with advantage.

References

Ajith Kumar, T.T., Kannan, R., Kannan, L., 1995. Remote sensing in mangrove research. Seshaiyana 3, 46–48.

Arnoff, S., 1982. Classification accuracy: a user approach. Photogramm. Engg. Remote Sensing 48(8), 1299–1307.

Green, E.P., Mumby, P.J., Edwards, A.J., Clark, C.O., 1996. A review of remote sensing for the assessment and management of tropical coastal resources. Int. J. Mar. Environ. 24, 1–40.

Krishnamoorthy, R., Remote Sensing for the assessment and management of mangroves: Indian Experience, (paper presented at the international workshop on "Application of Remote Sensing and GIS for sustainable Development at National Remote Sensing Agency," Hyderabad, 1997.

Navalgund, R.R. and Kiran Kumar, A.S., Indian Remote Sensing Satellite IRS-P4 Utilisation Plan, SAC-RSA/IRS-P4-UP/PP-02/97, Space Applications Centre (ISRO), Ahmedabad, India, May 1997.

Navalgund, R.R., Kiran Kumar, A.S., IRS P4 OCM Satellite: Space Applications Centre (ISRO), Ahmedabad, India.

Nayak, S.R., Gupta, M.C., Chawan, H.B., 1986. Wetland and shoreline mapping of the part of Gujarat coast using Landsat data, Scientific note (IRS-UP/SAC/MCE/SN06/86). Space Application Centre, Ahemdabad, p. 24.

Nayak, R.S., 1993. Role of Remote Sensing Application in the management of wetland ecosystems with special emphasis on Mangroves (Lecture delivered at the UNESCO Curriculum Workshop on "Management of mangrove Ecosystem and Coastal Ecosystem" at the Department of Marine Living Resource, Andhra University, Vishakhapatnam).

Oil Pollution and its Environmental Impact in the Arabian Gulf Region
M. Al-Azab, W. El-Shorbagy and S. Al-Ghais, editors

Chapter 5

Atmospheric-ocean prediction system for supporting oil spill monitoring: description and recent developments

S. Nickovic[1,*], V. Djurdjevic[2], B. Telenta[1], S. Music[1], B. Rajkovic[2] and A. Micallef[1]

[1]Euro-Mediterranean Centre on Insular Coastal Dynamics (ICoD), University of Malta, Malta; now at: Environment Division, Atmospheric Research and Environment Program, World Meteorological Organization, Geneva, Switzerland
[2]Faculty of Physics, Belgrade University, Belgrade, Serbia and Montenegro

Abstract

This chapter describes the atmospheric-marine system Regional Earth Observation Application for Mediterranean Sea Emergency Surveillance designed to provide necessary marine and weather predictions useful for oil spill monitoring. The system has been operationally implemented over the Mediterranean. The chapter presents recent extensions of the system related to a full two-way interaction model coupling. A new oil spill model being under development which is fully embedded into the ocean model is also described. Preliminary results of a hypothetical oil spill release simulation in the Persian Gulf are finally shown.

1. Modelling in RAMSES

1.1. Introduction

Due to frequent ship traffic and oil exploitation activities, regional natural environments such the Mediterranean and the Persian Gulf are exposed to high risk of oil spilling and ecological pollution. An accurate real-time prediction of atmospheric and marine conditions at the air–sea interface has been recognized as highly important tool in order to mitigate possible consequences of accidental and/or purposeful oil spilling. To perform such forecasts, advanced mathematical models are requested.

The Regional Earth Observation Application for Mediterranean Sea Emergency Surveillance (RAMSES) project funded by European Union (1998–2000) has been established following needs for oil spill (OS) monitoring in the Mediterranean. An effort to integrate work of remote sensing and environmental modelling of relevant processes has included the following project partners: European Space Agency (ESA/ESRIN, Italy),

*Corresponding author. *E-mail address:* snickovic@wmo.int (S. Nickovic).

Matra S&I (France), ACS (Italy), ICoD (Malta), Eurimage (Italy), FMA (Italy), SPOT Image (France), CEDRE (France), NARSS (Egypt) and CRTS (Morocco). This chapter describes concepts and results of ICoD that relate to marine and atmospheric modelling. Predecessors of the RAMSES project had involved only remote satellite sensing which appeared to be insufficient to fully describe environmental conditions responsible for driving the oil spill process. Within RAMSES, the new component – ocean and weather forecasting has been added to improve monitoring activities. This was one of the first successful attempts to routinely run coupled atmospheric–ocean models in the Mediterranean.

The RAMSES prediction system consists of the three major components: atmospheric, ocean and wave models. The regional atmospheric National Centers for Environmental Prediction (NCEP)/Eta model (Bryan, 1969; Janjic, 1977, 1979, 1984, 1990, 1994, 1996b, 1997, 2002, 2003; Mesinger et al., 1988; Janjic et al., 2001; Janjic 1996a) has a role to drive the other two components, the Princeton Ocean Model (POM) (Mellor, 1973; Mellor and Yamada, 1974, 1982; Blumberg and Mellor, 1982) and Wave model – WAM (WAMDI Group, 1988). Integration of these models within RAMSES is described in more details in Telenta et al. 2001.

1.2. Models

1.2.1. NCEP/Eta model

NCEP/Eta is the state-of-the-art model used for operational weather forecasts in the United States applied for research and operational purposes in many institutions worldwide. This is a regional grid-point model with advanced dynamics and physics incorporated. To increase the computational efficiency, a transformed latitude–longitude coordinate system is used in the model with a rotated coordinates having the origin at the model centre. In this way, strong meridian convergence is avoided at the northern and southern lateral model boundaries and longer model iteration time steps could be used. The model computational points in the horizontal are distributed on the semi-staggered Arakawa E grid in which scalar and vector variables are staggered in the horizontal. The choice of the grid provides appropriate calculation of the geostrophic adjustment processes in the atmosphere (gravity and inertia waves). Efficient explicit time integration numerical schemes are applied to keep computational time reasonably short. For numerical solving of the non-linear horizontal advection process, a method that preserves energy and squared vorticity and controls non-linear energy cascade is applied. The model vertical coordinate (called "eta") generates quasi-horizontal model levels. The model topography is composed of elementary topography grid point boxes with a step-like shape of the mountains.

Processes with scales smaller than the model grid resolution are parameterized, i.e. expressed in terms of the model grid-point parameters. So-called physical parameterization package of the model includes surface processes, turbulent mixing, convection, large-scale precipitation, lateral diffusion and radiation. This model component is particularly important since it directly interacts in coupling with other models. For parameterization of land surface processes, the Oregon State scheme is applied. Turbulence in the free atmosphere is parameterized using the Mellor–Yamada 2.5 closure level model (Mellor and

Yamada, 1974, 1982). Surface turbulent fluxes are the lower boundary condition for the turbulent scheme. Fluxes are calculated following the Monin-Obukhov theory (Chen et al., 1997). In a thin air layer close to the surface, there is not enough room to have well developed turbulent motion, and thus mixing on molecular level is dominant. To parameterize this process, Janjic (1994) have introduced a viscous sub-layer scheme. Over the ocean, the viscous sub-layer operates over three turbulent regimes, depending on values of the Reynolds friction number. The first regime relates to conditions under which molecular mixing dominates over the turbulent; the second one is the mixture of both mixing processes, the third regime is characterized with a well-developed turbulence. The third regime is connected to rough sea conditions. Parameterization of convection processes is based on the Bets–Miller–Janjic scheme. Convection is split into deep and shallow components with precipitation permitted to occur only in the former one. Large-scale precipitation is calculated according to values of predicted relative humidity. The radiation package is based on the Geophysical Fluid Dynamics Laboratory (Princeton) scheme, which includes interactive cloud overlapping, and influence of ozone and carbon dioxide on radiation absorption. The incoming short- and long-range fluxes are calculated at the air–sea interface and used for the ocean model forcing through the model-coupling scheme.

1.2.2. POM model

POM has been developed to simulate and/or predict ocean dynamic and physical processes. This is a primitive equation model with grid point method applied for solving partial differential equations. The basic equations have been defined in the terrain-following sigma coordinate system (Blumberg and Mellor, 1982) and using fully staggered C Arakawa horizontal grid. The model equations are written in a flux form. It is desirable in terms of computer economy to separate the vertically integrated equations (external mode) from the vertical structure equations (internal mode). This technique, known as mode splitting permits the calculation of the free surface elevation with little sacrifice in computational time by solving the velocity transport separately from the three-dimensional calculation of the velocity and the thermodynamic properties. A free sea-surface upper-boundary condition is imposed, which provides split of the model dynamics to internal and external wave dynamics. The leapfrog two-level time scheme is used for the model integration in time. The horizontal advection calculations are based on the Arakawa energy and enstrophy-conserving scheme. As in the case of the NCEP/Eta model, the Mellor–Yamada scheme is used to describe turbulent processes. This method permits generation of a well-mixed surface layer what is particularly important for coastline and shallow sea dynamics. Free see surface in the model permits implementation of the tidal effects. The radiation physics includes effects of the transferred short-wave radiation through the water. The Smagorinsky diffusivity scheme for horizontal diffusion with a constant or biharmonic diffusion has been used.

1.2.3. WAM model

The WAM model solves the evolution of a two-dimensional ocean wave spectrum. In contrast to the first- and second-generation model, this model version introduces no assumption on the spectral shape. It computes two-dimensional wave variance spectrum through integration of the governing transport wave equation. The equation includes advection and wave sources. The source function is composed of the wind input term and the dissipation term. The surface waves extract momentum from the airflow and thus the stress in the surface layer depends on both wind speed and wave-induced stress. The growth rate of the waves depends on the friction velocity and the roughness length of the sea surface.

1.3. Operations

RAMSES is as a fully automatic modelling system: from the phase of collecting necessary data input, through preparing data and running models, to the final phase of preparing results in graphical and binary forms and submitting forecasting maps to a dedicated web site. RAMSES uses two alternative sources of input atmospheric data – the meteorological global forecasts from either the NCEP or the European Centre for Medium-Range Forecasts (ECMWF). In order to specify lower boundary conditions of the atmospheric model, high-resolution information on vegetation cover and soil types have also been used.

Global weather forecast data were used to define daily-updated initial and boundary conditions for the regional (Mediterranean) NCEP/Eta model. After data conversion to the model grid and with run time/space parameters defining the model domain, geometry, resolutions, time step and other parameters and constants. The model was executing once per day over the 3-day and 1.5-day periods with horizontal resolution of 16 km and 4 km for the regional and local domains, respectively.

The models produce a number of parameters: sea and near sea surface parameters – currents, air and sea temperature, direction and significant height of wind waves, wind, air moisture, air momentum, surface fluxes of heat, momentum, moisture, radiation (short- and long-wave) deep sea parameters – currents, salinity, temperature; conventional upper air parameters – temperature, geopotential, wind, precipitation and cloudiness. Figure 1 shows an example of the RAMSES operations provided daily to the users. This is 36-hour forecast of 10 m wind performed over the Central Mediterranean, demonstrating how the high-resolution (4 km) atmospheric model represents near-surface airflow conditions.

Within the RAMSES activities, validation of different forecasting products has been performed. Examples of comparisons of the WAM wave height and the NCEP/Eta 10 m wind against observations is shown in Figure 2a,b, respectively.

The modelling system was designed to be fully modular and automatic. It can be set up and compiled in a short time over any geographical region. Implementation of the whole modelling software requires low cost investments. Minimum technical requirements to setup the models operationally are: 128 Kbps Internet connection, 800 MHz Pentium computer, 128 Mb RAM and 20 Gb disk space. The software has been developed under the Linux/Unix operating system. No licenses or charges are required for input data (NCEP, Washington data), software and graphical packages.

1.4. Model interactions

The integration of NCEP/Eta with the other two models was based on a one-way forcing method. In the post-processing phase, the regional model output was prepared to provide

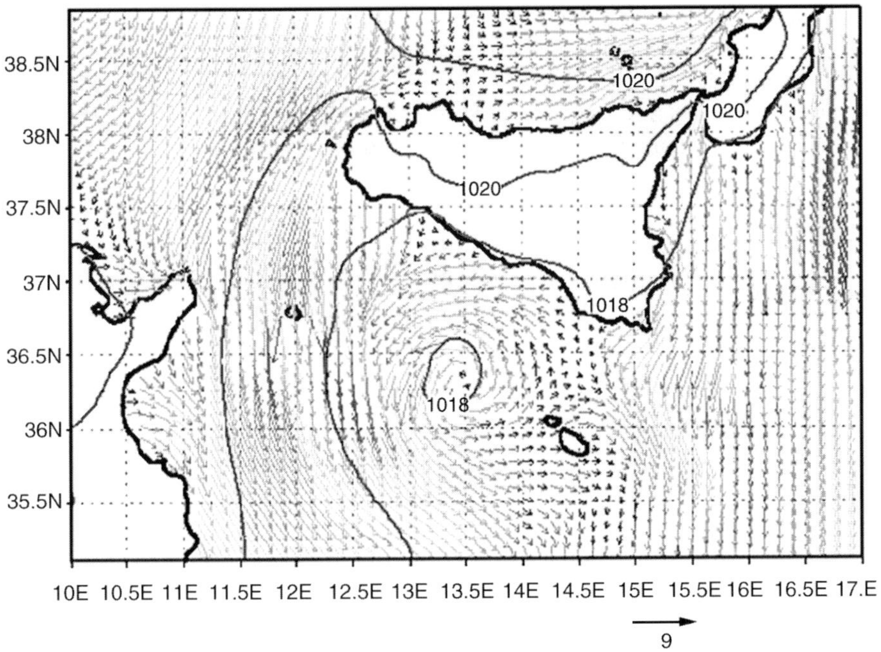

Figure 1. 10 m wind conditions as predicted by the NCEP/Eta model. The length of arrows indicates the wind intensity.

forcing of the POM and WAM regional models with frequency of the output of 3 h. Once the daily run cycle of NCEP/Eta, POM and WAM were finished, output of these models was used to perform another level of model integration – to nest a number of local high-resolution models into their regional parents. Nesting has been done by driving three different models through the lateral boundary conditions. Local model runs were performed over the periods of one-and-a-half-day periods with the horizontal resolution of 4 km. Driving process includes supplying of the NCEP/Eta surface fluxes and momentum

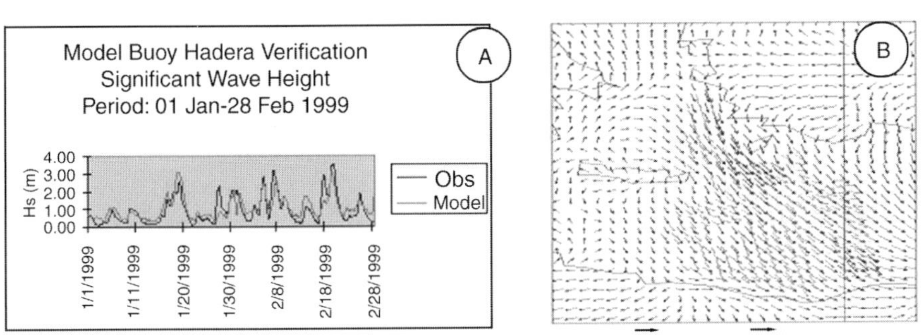

Figure 2. (A) Comparison between model predicted and observed values of the significant wave height at Hadera (Israel); (B) comparison between model predicted 10 m wind and surface wind observed by ERS-2 satellite (ERS-2 wind are coloured, and NCEP/Eta wind are black arrows).

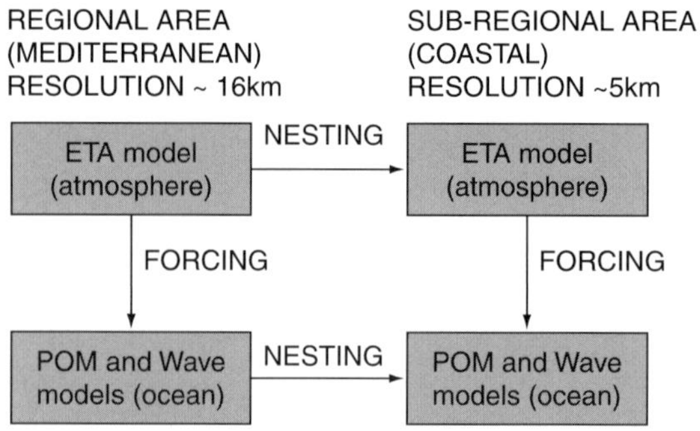

Figure 3. Model interconnection through forcing and nesting in the RAMSES system.

at the air–sea interface. In addition of using regional models with a coarser horizontal resolution over the whole Mediterranean, a series of small-domain high-resolution models is nested to the coarse models over several local regions having high risk of oil spilling. Schematic representation of the model forcing and nesting is shown in Figure 3; domains of regional and local models applied in RAMSES are shown in Figure 4.

2. Further developments: fully integrated models

Interaction of the atmospheric and marine environments is the process done in two directions. Part of the atmospheric energy is transferred to the ocean providing forcing of the marine processes. Simultaneously, the ocean is a source of energy for the atmosphere. This interaction mechanism is maintained through the momentum, heat and moisture flux exchange at the air–sea interface.

Due to complexity of the feedback process, most of today's integrated atmosphere–marine modelling systems are based only on a one-way forcing approach in which these two components are run off-line. The RAMSES system has been setup to have a one-way off-line forcing approach as well. A common practice today is that oil spill models available in the commercial market and in the scientific community are off-line driven by

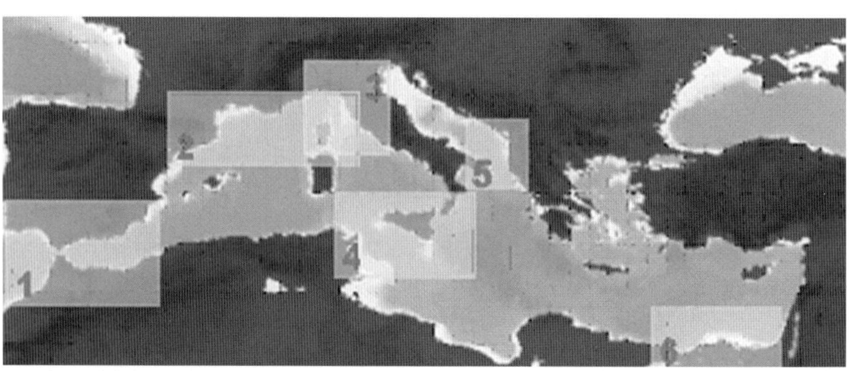

Figure 4. Regional and local model domains in RAMSES.

the wind forecasts provided by atmospheric models. In this way, useful information on dynamics and physics of the ocean models is completely neglected. For example, ocean model information such as turbulent mixing, vertical transport, salinity and temperature conditions. In this way, their well-resolved ocean spatial and temporal variations are not incorporated into off-line driven oil spill models. An oil spill model fully embedded into an ocean model would be an appropriate and consistent solution. Therefore, a more complete integration of different natural environments and systems such as atmosphere, ocean and pollution would be more useful in achieving further improvement of routine prediction of the oil spill fate.

It is envisaged that there are at least two components in modelling systems such as RAMSES that could be developed to adequately represent/predict the oil spill process. *One* is replacement of the one-way forcing by integrating the atmospheric and ocean models through two-way on-line interaction. This upgrade should improve the overall quality of both weather and marine forecasts. The *second* aspect is developing a modelling system in which the oil spill model is driven on-line and fully embedded into the ocean model. In this way, all information from the ocean model relevant for the oil spill process would be used simultaneously during the course of the coupled models execution. Such new approach is expected to considerably refine predictions of the oil spill process.

2.1. Fully coupled NCEP/Eta–POM system

Following the above concept of integrated models, Djurdjevic and Rajkovic (2002) in his recent study have successfully developed a fully integrated NCEP/Eta–POM modelling system. Within each time step of the model physics, the NCEP/Eta model continuously calculates all driving fluxes required by the POM model. At the end of each such time step, the POM model is called as a program routine to perform its calculations with the atmospheric updates. Once this step is finished, the ocean model supplies the new values of the sea surface temperatures used now by the Eta model as an updated lower boundary condition. Such interaction continues up to the end of the models simulation period.

The integrated system has been tested in model executions over the extended time period. The modelling system has been run from 1st December 2001 to 30th June 2002 over the Mediterranean region with horizontal resolution of 0.25°. The atmospheric conditions have been updated every 30 days with observed values. The ocean model has been initialized using MODB ocean climatology for the Mediterranean; monthly updates are done using simulated ocean conditions from the previous month. The coupled system shows substantial level of stability of SST and other parameters on both sides of the air–sea interface. At the same time, when compared against observed one (Fig. 5), simulated SST shows a considerable agreement over the whole period of the experiment. This has been an important severe test of the modelling system performances representing a precondition for possible operational applications.

2.2. New integrated NCEP/Eta–POM–oil spill modelling system

An attempt has been made to develop an oil spill (OS) model fully integrated with its driver, the POM ocean model, which is at the same time integrated with the NCEP/Eta

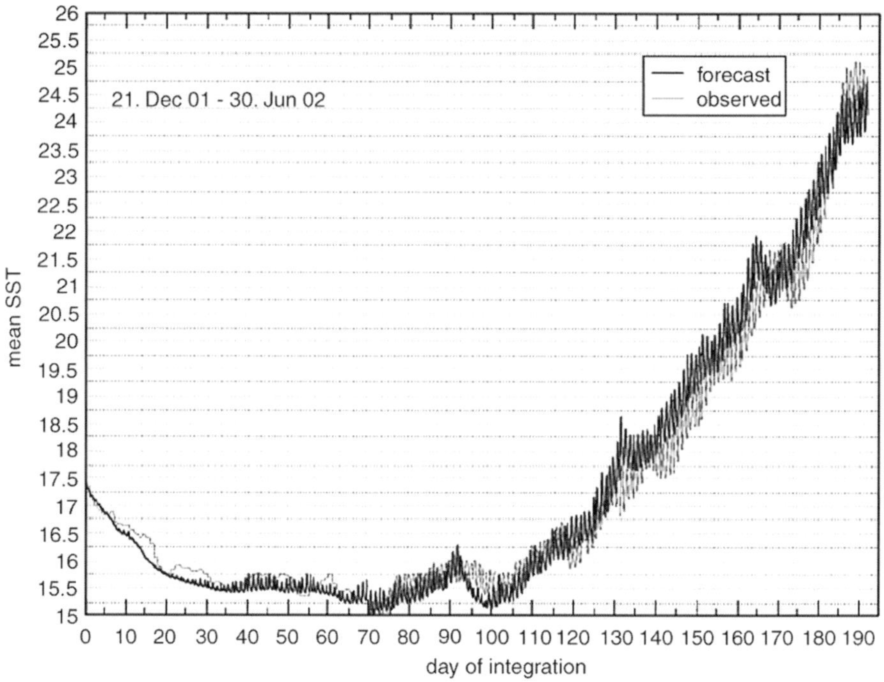

Figure 5. Fully integrated atmospheric–ocean modelling system: simulated vs. observer spatial mean of sea surface temperature.

atmospheric model. In our approach, a governing continuity equation for spilled oil has been introduced as additional equation in the POM model. The continuity equation is of the Euler type in which oil concentration is driven by the ocean model variables. The equation is represented by

$$\frac{\partial C}{\partial t} = -u^* \frac{\partial C}{\partial x} - v^* \frac{\partial C}{\partial y} - (w^* - v_g)\frac{\partial C}{\partial z} - \nabla(K_H^* \nabla C) - \frac{\partial}{\partial z}\left(K_Z^* \frac{\partial C}{\partial z}\right)$$
$$+\left(\frac{\partial C}{\partial t}\right)_{SRC} - \left(\frac{\partial C}{\partial t}\right)_{SNK}$$

where u, v and w are horizontal and vertical sea current components, K_H and K_Z are lateral and vertical diffusion coefficients, v_g is deposition velocity and C is oil concentration. The superscript ($*$) indicates ocean model variables. SRC and SNK terms describe sources and sinks of spilled oil. Being yet in its initial stage of development, the new oil spill model currently includes horizontal advection and diffusion terms. In the next phase, other processes relevant for oil spill fate such as vertical mixing and advection, coagulation, sedimentation biodegradation, evaporation, emulsification, dissolution and oxidation will be parameterized.

The integrated NCEP/Eta–POM–OS modelling system has been run over the Persian Gulf to demonstrate capabilities of the new approach in integrating three environmental

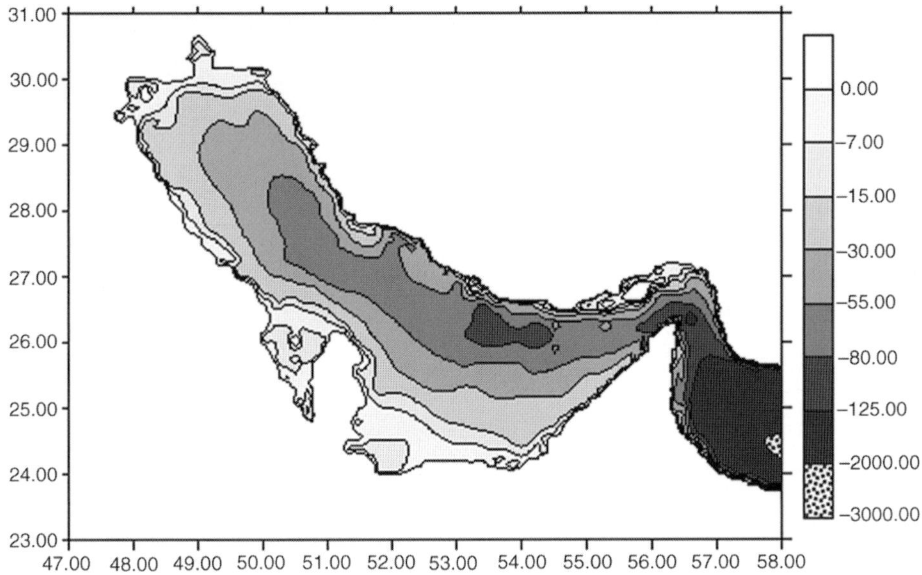

Figure 6. Domain of the regional POM model and its bathimetry.

models and simulating oil spill evolution. Real-time atmospheric data are used to initialize and drive the atmospheric model over the period of February 2002. Observed profiles of salinity and ocean temperature are used for POM initialization. Regional POM and NCEP/Eta have been run with resolution of 0.25°. Figure 6 shows bathymetry in POM and the domain of regional NCEP/Eta and POM models in the experiment. Execution of a nested POM–OS system with resolution of 0.004° located over United Arab Emirate coastal waters has been then performed. Simulated sea surface temperature and currents valid at 12:00 13th February 2002 are shown in Figure 7.

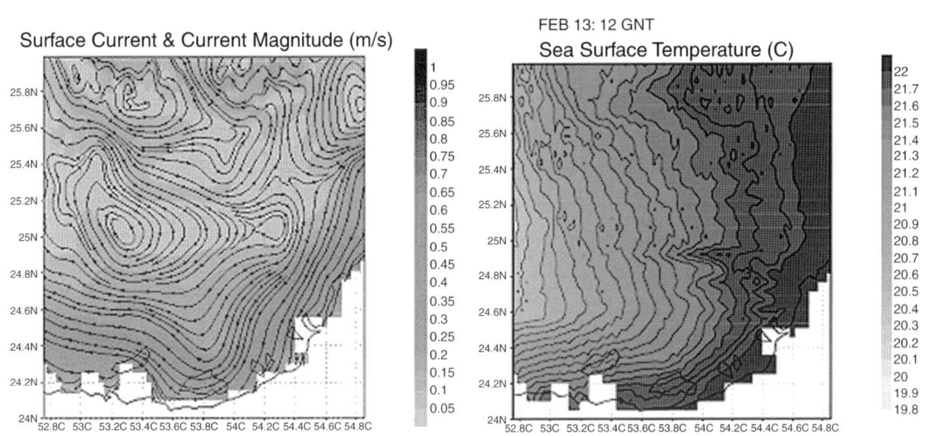

Figure 7. Simulated sea surface currents (above) and temperatures (below) in the United Arab Emirates waters.

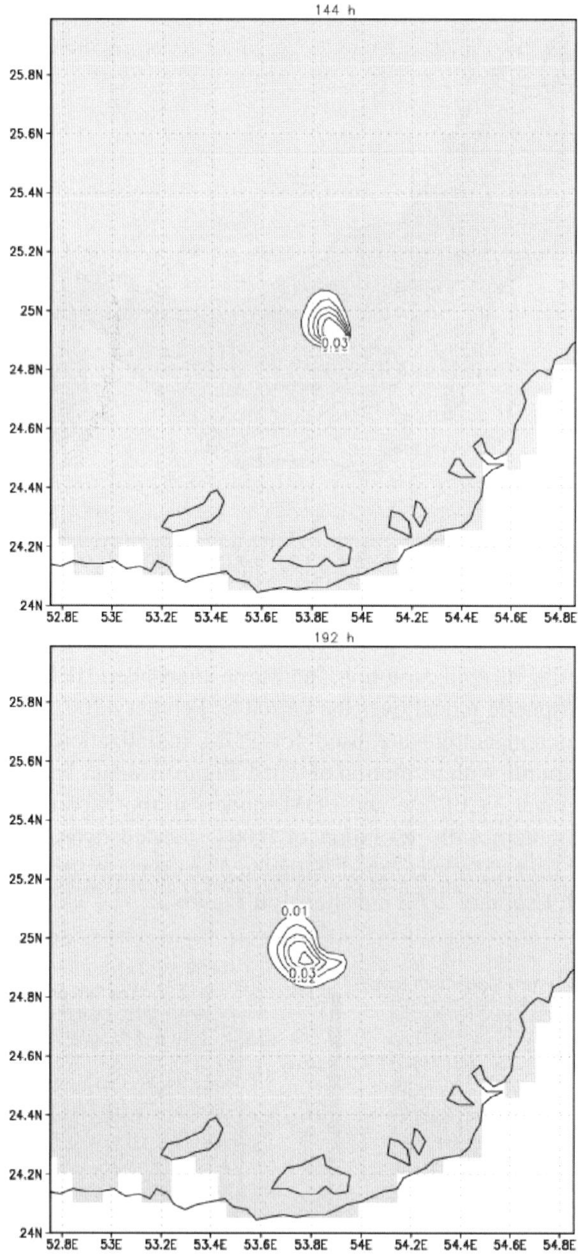

Figure 8. Hypothetical oil spill simulated in the United Arab Emirates waters after 144 (top) and 192 (bottom) hours of model integration.

During the course of the NCEP/Eta–POM–OS execution, an arbitrary mass of oil has been released into the ocean during initial eight consecutive days of the experiment. The oil spill dynamical evolution as shown in Figure 8 is a result of full interaction of three integrated models. Although the OS model currently includes only limited number of processes relevant for the oil spill fate, the experiment demonstrates how OS dynamics is affected by the integrated NCEP–POM–OS models.

3. Conclusions

Mathematical modelling of air–sea interactions is important and challenging scientific task yet not satisfactory represented. Within the EU-funded RAMSES project a successful integration of the atmospheric and ocean model has resulted in establishment of an operational forecasting system designed to support oil spill monitoring activities in the Mediterranean. However, RAMSES model integration has been based on the one-way model forcing concept. It also has not incorporated oils spill modelling.

In a recent effort, the atmospheric and ocean models have been fully coupled through a two-way interaction method. In addition, a simplified oil spill model has been driven by and fully embedded into the ocean model structure. In the future work, the oil spill model will be completed with missing physical and dynamic parameterizations.

The oil spill dispersion is rather complex process dependent on both atmospheric and marine conditions. We thus envisage that full integration of atmospheric, ocean and oil spill models is the most promising approach to incorporate all major environmental influences on oil and to achieve the goal of accurate prediction of the oil spill fate in accidental situations.

References

Blumberg, A.F., Mellor, G.L., 1982. A description of a three-dimensional coastal ocean circulation model. In: Heaps, N. (Ed.), Three-Dimensional Coastal Ocean Models, Vol. 4. American Geophysical Union, Washington, DC, p. 208.

Bryan, K., 1969. A numerical method for the study of the circulation of the world ocean. J. Comp. Phys. 4, 347–376.

Chen, F., Janjic, Z., Mitchell, K., 1997. Impact of atmospheric surface-layer parameterization in the new land-surface scheme of the NCEP mesoscale Eta model. Bound. Lay. Meteorol. 48, 391–421.

Djurdjevic, V., Rajkovic, B., 2002. Air–sea interaction in the Mediterranean Sea. In: Spring Colloquium on the Physics of Weather and Climate ICTP, Trieste, Italy.

Janjic, Z.I., 1996a. The Mellor–Yamada level 2.5 scheme in the NCEP Eta model. 11th Conference on NWP, Norfolk, VA, American Meteorological Society, pp. 333–334.

Janjic, Z.I., 1977. Pressure gradient force and advection scheme used for forecasting with steep and small scale topography. Contrib. Atmos. Phys. 50, 186–199.

Janjic, Z.I., 1979. Forward–backward scheme modified to prevent two-grid-interval noise and its application in sigma coordinate models. Contrib. Atmos. Phys. 52, 69–84.

Janjic, Z.I., 1984. Non-linear advection schemes and energy cascade on semi-staggered grids. Mon. Weather Rev. 112, 1234–1245.

Janjic, Z.I., 1990. The step-mountain coordinate: physical package. Mon. Weather Rev. 118, 1429–1443.

Janjic, Z.I., 1994. The step-mountain Eta coordinate model: further developments of the convection, viscous sublayer and turbulence closure schemes. Mon. Weather Rev. 122, 927–945.

Janjic, Z.I., 1996b. The surface layer in the NCEP Eta model. 11th Conference on NWP, Norfolk, VA, American Meteorological Society, pp. 354–355.

Janjic, Z.I., 1997. Advection scheme for passive substances in the NCEP Eta model, Research Activities in Atmospheric and Oceanic Modelling, WMO, Geneva, CAS/JSC WGNE, 3.14.

Janjic, Z.I., 2002. Nonsingular Implementation of the Mellor–Yamada Level 2.5 Scheme in the NCEP Meso model. NCEP Office Note No. 437, p. 61.
Janjic, Z.I., 2003. A nonhydrostatic model based on a new approach. Meteorol. Atmos. Phys. 82, 271–285.
Janjic, Z.I., Gerrity, J.P. Jr, Nickovic, S., 2001. An alternative approach to nonhydrostatic modeling. Mon. Weather Rev. 129, 1164–1178.
Mellor, G.L., 1973. Analytic prediction of stratified planetary surface layers. J. Atmos. Sci. 30, 1061–1069.
Mellor, G.L., Yamada, T., 1974. A hierarchy of turbulence closure models for planetary boundary layers. J. Atmos. Sci. 31, 1791–1806.
Mellor, G.L., Yamada, T., 1982. Development of a turbulence closure model for geophysical fluid problems. Rev. Geophys. Space Phys. 20, 851–875.
Mesinger, F., Janjic, Z.I., Nickovic, S., Gavrilov, D., Deaven, D.G., 1988. The step-mountain coordinate: model description and performance for cases of Alpine lee cyclogenesis and for a case of an Appalachian redevelopment. Mon. Weather Rev. 116, 1493–1518.
Telenta, B., Nickovic, S., Music, S., Ryabinin, V., 2001. Operational system for marine prediction in the Mediterranean: description and products. Proceeding of EuroGOOS Conference.
WAMDI Group, 1988G. The WAM model – a third generation ocean wave prediction model. J. Phys. Oceanogr. 18, 1775–1810.

Oil Pollution and its Environmental Impact in the Arabian Gulf Region
M. Al-Azab, W. El-Shorbagy and S. Al-Ghais, editors

Chapter 6

Applications of denaturing gradient gel electrophoresis (DGGE) and microsensor techniques in oil biodegradation studies

Raeid M.M. Abed* and Stefan Grötzschel

Max-Planck Institute for Marine Microbiology, Celsiusstrasse 1, D-28359 Bremen, Germany

Abstract

Denaturing gradient gel electrophoresis (DGGE) and microsensor techniques have been widely used in microbial ecology. In this review, we focus on the applications of these techniques in the context of oil biodegradation. DGGE provides information on microbial diversity of polluted ecosystems. Furthermore, using this technique, changes in the structure of microbial communities over time could be monitored. Microsensors detect the microenvironment of contaminated sediments and monitor changes in various microbial metabolic processes such as respiration, photosynthesis and sulfate reduction. The application of microsensors in oil biodegradation studies has been restricted to cyanobacterial mats. The role of these mats in the breakdown of petroleum compounds has been recently realized following the Gulf War in 1991. Therefore, some of our findings in this field are reviewed.

1. Introduction

Oil pollution poses a serious threat for terrestrial and marine ecosystems. Huge oil spills not only have dramatic effects on marine life but also result in significant contamination of shorelines (Swannell et al., 1996). Several physical and chemical methods have been implemented in order to clean up contaminated sites. Bioremediation, defined as the intentional use of biodegradation processes to eliminate environmental pollutants (Madsen, 1998), has been widely considered (Swannell et al., 1996; Vogel, 1996; Jones, 1998; Head and Swannell, 1999; Timmis and Pieper, 1999; Samanta et al., 2002). Considerable attention has been given to the identification of hydrocarbon-oxidizing bacteria in natural environments because of the possibility of utilizing their biodegradation potential in the treatment of oil spills. The significance of photosynthetic bacterial communities such as cyanobacterial mats in bioremediation was first realized after the Gulf War in 1991 (Sorkhoh et al., 1992). Following the oil spill, microbial mats, dominated by cyanobacteria, colonized the polluted sediments (Hoffmann, 1996; Höpner

*Corresponding author. *E-mail address:* rabed@mpi-bremen.de (R.M.M. Abed).

et al., 1996). The presence of these cyanobacterial mats was remarkably associated with oil pollution, so that oil-free sediments were free of such mats. This kind of mats had not been previously reported in the region, except the unpolluted Abu Dhabi mats (Golubic, 1992). A similar phenomenon was observed in the highly polluted Wadi Gaza (Gaza Strip-Palestine) on the Mediterranean coast. This stream receives a variety of pollutants such as diesel oil, hydrocarbons, sewage, pesticides, solid waste as well as agricultural and industrial discharges. In spite of this high level of pollution, cyanobacterial mats have also developed there. Since cyanobacterial mats are evidently pollution resistant, it is assumed that they contain communities that possess oil-degrading ability. This has stimulated our research group to explore the role of these cyanobacteria-dominated communities in oil biodegradation and to investigate their possible utilization for bioremediation purposes. Some of our findings will be reviewed here.

Most biodegradation studies have been performed on isolated oil-degrading strains or in the laboratory under simulated field conditions. However, the inability to duplicate natural conditions in the laboratory often leads to the development of bacterial communities different from those present under *in situ* conditions. Furthermore, the isolated strains may represent an insignificant proportion of all microorganisms thriving under natural conditions since the isolates are often selected by the supplied culture conditions. Identification of bacteria in the field by direct microscopy is very difficult due to their simple morphology and lack of conspicuous traits that distinguish them from each other. Therefore, there is a need for techniques enabling us to identify important microorganisms under *in situ* conditions and to monitor changes in community structure and function. For that purpose, new molecular and microsensor techniques have been introduced to the field of molecular ecology. Denaturing gradient gel electrophoresis (DGGE) is a culture-independent fingerprinting technique which has been widely used to study microbial diversity and community changes over time or after disturbances (Muyzer et al., 1993; Muyzer and Smalla, 1998). Microsensors are glass electrodes which are used to measure various parameters related to microbial activity under *in situ* conditions. DGGE and microsensor techniques have been applied to many environmental systems in order to resolve various ecological questions. However, their use in oil biodegradation studies has been relatively limited.

In this review, we discuss the applications of DGGE and microsensor tools in studying the structure and function of microbial communities at oil polluted sites. We additionally report on our research on cyanobacterial mats and their role in oil biodegradation using these techniques. The aim of this review is to provide researchers with extra tools for investigating processes that take place in the field and to find out the role of bacteria in these processes.

2. Denaturing gradient gel electrophoresis

DGGE was first introduced by Muyzer et al. in 1993 as a method for profiling complex environmental microbial populations. Since then, this technique has had many applications in microbial ecology. A detailed overview of such applications is given in Muyzer and Smalla (1998). In the present review, we will only emphasize on its applications in biodegradation studies.

2.1. How does DGGE work?

The theory behind the separation is relatively straightforward (Fig. 1). DNA is a double-stranded polymer in which the two strands are held together by hydrogen bonds between complementary bases of opposite strands. As one increases the concentration of certain destabilizing chemicals, these interactions become weak, and the two strands separate (the DNA "melts"). The concentration of denaturants at which this occurs depends the DNA sequence. Stretches rich in G and C (which pair with three hydrogen bonds) are more stable than stretches rich in A and T (which pair with only two hydrogen bonds). Identical DNA molecules which differ in a single base within a "weak" domain will have different overall melting temperatures. If electrophoretic separation is carried out through a stable, increasing gradient of chemical denaturants (formamide and urea are usually used), DNA molecules will be halted in their migration at the concentration at which the DNA strands dissociate. This is because single-stranded molecules become entangled in the gel matrix. Typically, in order to improve the sensitivity of electrophoretic separation, complete strand separation is prevented by the addition of a GC-rich stretch with very high melting domain (GC clamp) at one end of the molecule during polymerase chain reaction (PCR). This improves resolution when a difference in sequence occurs in a high-melting domain. For mixed (relatively complex) environmental samples of 16S rRNA gene fragments, each sample is run in a single lane, and the various alleles stop at different places within that lane, yielding a pattern of DNA bands which typifies the sample. In principle, each band represents an allele of the 16S rRNA gene, and thus a group of organisms that are closely related phylogenetically.

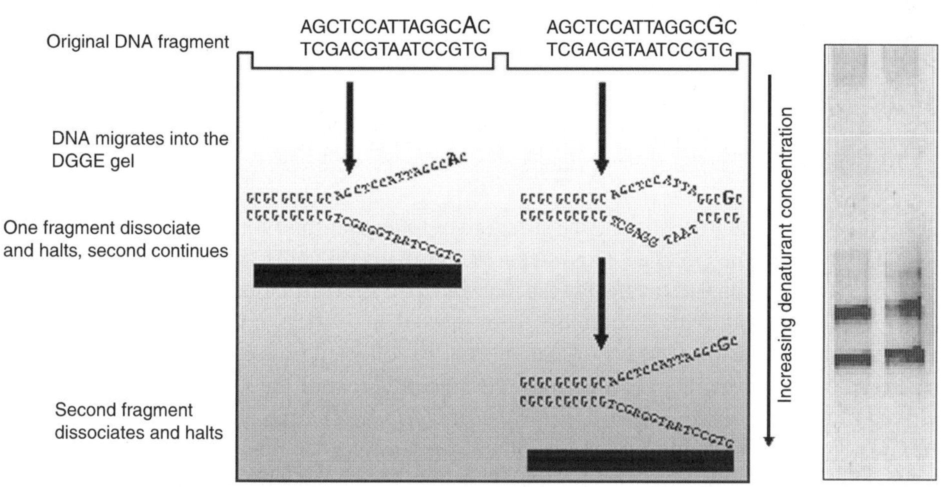

Figure 1. A schematic illustration showing the principle of DGGE technique. The example shows the separation of two 16S rRNA fragments with a single base pair difference. The right panel shows a DGGE gel exhibiting a community dominated by two populations.

2.2. Why 16S rRNA?

The 16S rRNA molecule is an excellent genetic tool for community analysis and phylogenetic comparison as it is present in all organisms in high numbers and it possesses both conserved and variable sequences which enables easy design of specific primers and probes for covering different phylogenetic ranges. Here, therefore, DGGE based on the separation of 16S rDNA genes is an excellent tool to study complex microbial communities.

2.3. What are the steps of community analysis by DGGE?

A scheme of different steps involved in community analysis by DGGE is presented in Figure 2. The first step is the extraction of total nucleic acids, DNA and RNA, from environmental samples. The extracts are then subjected to amplification by PCR using different sets of primers. Nübel et al. (1997) developed primers that specifically amplify cyanobacterial and plastidic 16S rRNA genes. Several other primers, which amplify the 16S rDNA of all microorganisms of the domain Eubacteria or specific subgroups of bacteria, are also available (Table 1). The PCR products are loaded on the DGGE gel and the electrophoresis is then performed at 60°C and a constant voltage of 200 V for 3.5 h. Different denaturant gradients, temperatures and running periods may be used for different primers. After electrophoresis, the various alleles of the 16S rRNA fragments amplified are subsequently separated on polyacrylamide gels. The banding pattern reflects the species composition of the environmental sample. The bands can then be excised from the gel, re-amplified with the same primers. Part of the amplification product is then run on a second DGGE gel to check that it forms a single band. Clean single bands are sequenced and compared with 16S rRNA sequences in the existing databases (currently about 50,000 sequences) to find the closest relatives. Detailed protocols of the above mentioned steps can be found in Muyzer et al. (1998).

2.4. Applications in oil biodegradation studies

2.4.1. Determination of bacterial diversity in oil polluted sites

DGGE has been used to determine the structure of total microbial communities or particular populations in various polluted ecosystems or following oil spills. Kasai et al. (2001) used DGGE to study the long-term effect of the *Nakhodka* oil spill on marine bacterial populations. This oil spill resulted in the contamination of more than 500 km of the Japan Sea coastline with more than 5000 ton of heavy oil. Samples were collected from contaminated sediments at different time intervals over a period of 2.5 years. Most of the DGGE bands identified derived from the *Cytophaga/Flavobacterium/Bacteriodes* phylum, the α-*Proteobacteria* or the Cyanobacteria. Community composition in the seawater samples was different from that in the oil paste samples. These communities included strains phylogenetically related to oil degraders such as *Sphingomonas subarctica* and *Alcanivorax borkumensis*.

In another investigation, DGGE was used to study the cold-adapted bacterial communities in petroleum contaminated sediments from two Canadian environments

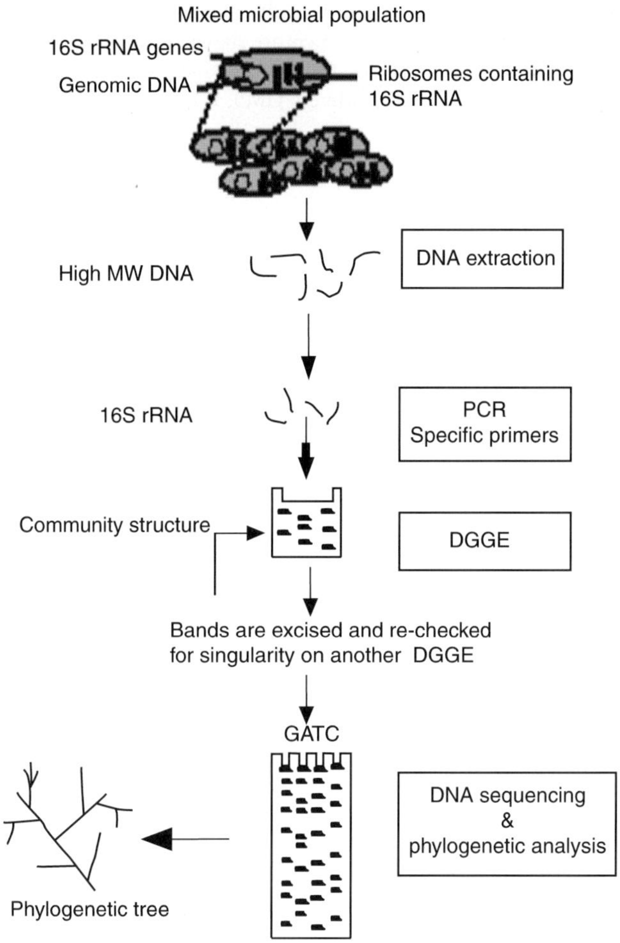

Figure 2. An overview of a microbial community study using denaturant gradient gel electrophoresis (DGGE). Nucleic acids are extracted from an environmental sample, amplified by polymerase chain reaction (PCR) and separated on a DGGE gel. The bands excised from the DGGE gel are sequenced and phylogenetically analysed.

(Juck et al., 2000). The analysis showed that, compared to pristine controls, bacterial diversity was lower at one contaminated site but the same or higher at the other. Phylogenetic analysis of the excised bands suggested that the community was dominated by high G + C Gram positive *Actinomycetales* (63.6%) and *Proteobacteria* (36.4%). The authors concluded that species diversity was determined by geographical origin of the samples rather than by the level of oil pollution.

Recently, this technique has been applied to photosynthetic communities to study their role in oil biodegradation by Abed et al. (2002). In this study, DGGE was used to explore the bacterial diversity of cyanobacteria-dominated microbial mats exposed to high levels of pollution from Wadi Gaza, Palestine. Two sets of primers were used, one specific for

Table 1. List of selected primers used to amplify small subunit ribosomal RNA used for DGGE. The target sites, sequences and specificity of the primers are shown. R (reverse) and F (forward) indicate the orientation of the primers. Y indicates C/T and W indicates A/T. A GC clamp (GC-rich sequence) is attached to the 5′-end of the primers GM5, CYA359, ARC344 and Eukaryotic F to prevent complete dissociation.

Primer	Target site	Sequence (5′ to 3′)	Specificity	Reference
GM5F	341–357	CCT ACG GGA GGC AGC AG	Bacteria	Muyzer et al. (1993)
907 RC	907–926	CCG TCA ATT CCT TTG AGT TT	Bacteria	Muyzer et al. (1998)
907 RM	907–926	CCG TCA ATT CMT TTG AGT TT	Bacteria	Muyzer et al. (1998)
CYA 359F	359–378	GGG GAA TYT TCC GCA ATG GG	Cyanobacteria	Nübel et al. (1997)
CYA 781 R	781–805	GAC TAC WGG GGT ATC TAA TCC CWTT	Cyanobacteria	Nübel et al. (1997)
Eukaryotic F	1427–1453	TCT GTG ATG CCC TTA GAT GTT CTG GG	Eukarya	Van Hannen et al. (1998)
Eukaryotic R	1616–1637	GCG GTG TGT ACA AAG GGC AGG G	Eukarya	Van Hannen et al. (1998)
ARC 344F	344–363	ACG GGG YGC AGC AGG CGC GA	Archaea	Casamayor et al. (2000)
915 R	915–934	GTC CTC CCC CGC CAA TTC CT	Archaea	Casamayor et al. (2000)

Cyanobacteria and the other targeting all bacteria. The cyanobacterial primers yielded primarily 16S rDNA sequences related to those from *Phormidium* and *Oscillatoria* species. This finding was in good agreement with microscopic observations. Bacteria belonging to the *Cytophaga–Flavobacterium–Bacteriodes* group, γ and β subclasses of the *Proteobacteria* and the green non-sulfur bacteria were also detected.

DGGE has also been used to study anaerobic microorganisms and their role in oil biodegradation. For example, Kleikemper et al. (2002) investigated the response of sulfate-reducing bacteria in a petroleum hydrocarbon contaminated aquifer to the addition of different carbon sources. In all treatments, sulfate reduction rates were stimulated following carbon source addition, whereas DGGE profiles indicated no changes in the community structure. Watanabe et al. (2000) used DGGE to investigate the bacterial populations in petroleum contaminated groundwater. Their results suggested that novel members of the epsilon subclass of the *Proteobacteria* were major populations in the petroleum-contaminated cavity groundwater.

2.4.2. Population dynamics

Addition of oil induces dramatic changes in the composition of microbial communities, which can be monitored by DGGE. For example, Macnaughton et al. (1999) followed changes in the bacterial community following a simulated coastal oil spill. The DGGE patterns showed no significant changes in uncontaminated controls, while the bacterial community in the oiled plots apparently underwent substantial shifts. The community composition was different when nutrients were added in addition to oil. Oil treatment was found to stimulate the growth of gram negative microorganisms within the α-*Proteobacteria* and *Cytophaga–Flavobacterium–Bacteriodes* phyla. α-*Proteobacteria* were never detected in uncontaminated controls, suggesting a possible involvement of this group in the degradation of oil. In a study by Duarte et al. (2001), sediment samples were exposed to varying concentrations of sulfurous oil. The number of DGGE bands decreased with increasing oil concentration, indicating the toxic effect of the oil. Sequence analysis showed that the organisms present at high oil concentration were related to *Actinomycetes* sp. and *Arthrobacter* sp. as well as an unknown bacterium.

Monitoring DGGE patterns in biodegradation studies may provide hints of who is (are) mainly responsible for the observed breakdown of oil compounds. For example, Abed et al. (2002) demonstrated the ability of cyanobacterial mats originating from Wadi Gaza to degrade phenanthrene and dibenzothiophene completely and pristane and *n*-octadecane partially within 3 days of incubation. It was concluded that members of the group *Holophaga–Geothrix–Acidobacterium* phylum may be responsible for the degradation of the four model compounds. The DGGE patterns showed that members of this group were apparently enriched in the presence of the model compounds, both in the light and in the dark, but did not appear in any of the controls. The analysis also indicated a possible involvement of green non-sulfur bacteria in the biodegradation process. Cyanobacteria seem to play a trivial role in the breakdown of petroleum compounds by applying oxygen for the aerobic heterotrophic degraders. However, it should be kept in mind that such conclusions based on DGGE analysis remain speculative unless proven by isolation of the respective organisms.

In some cases, the community may not undergo changes even when oil components are added. An example is the study by Grötzschel et al. (2002b), in which mats were incubated for 18 weeks in the presence of oil compounds (Fig. 3, left). In spite of the observed oil biodegradation and the long incubation period, no changes in community composition were detected by DGGE analysis.

2.4.3. Bioremediation studies

Bioremediation can be achieved either by bioaugmentation or biostimulation. Bioaugmentation, otherwise known as seeding involves the addition of oil-degrading bacteria to supplement the existing microbial communities. Biostimulation involves the addition of nutrients or other fertilizing cosubstrates to stimulate the growth of indigenous oil degraders. It is essential to follow the fate of added microorganisms in the former approach and the impact of the added stimulators on the present community in the latter. Ogino et al. (2001) used DGGE to follow microbial succession during a biostimulation process. They found that the community structure was disrupted by the biostimulation treatment but recovered immediately after the end of fertilization. The impact of different levels of inorganic nutrients on biodegradation rates and on bacterial community composition was studied by Roling et al. (2002). Oil biodegradation was stimulated upon addition of nutrients, irrespective of their concentration. Different microbial communities were apparently selected in all treatments, but this could not be correlated with nutrient levels.

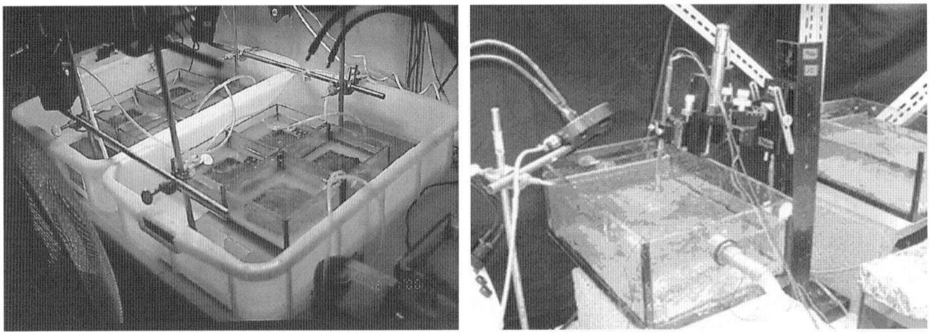

Figure 3. Left, biodegradation experiment with intact cyanobacterial mat pieces from a hypersaline environment. The mats were incubated with four petroleum compounds immobilized on clay particles. Right, setup to measure O_2 profiles using oxygen microsensor in a microbial mat incubated in a glass aquarium.

It was concluded that bioremediation treatments dramatically reduced bacterial diversity but this was accounted for by the selection of oil degrading bacteria.

2.4.4. Monitoring of enrichment cultures: guided, rather than random, isolation

It has been generally considered that functionally relevant populations in the environment are difficult to isolate. Many of the strains isolated by enrichment cultivation have been found to represent rare populations in the environment. DGGE can be used for the screening of isolates that have the dominant phylogenetic signatures of the original ecosystems. Several studies have demonstrated the use of DGGE to monitor the development of oil-degrading enrichment cultures (Friedrich et al., 2000; Wilson et al., 2001; Frontera-Suau et al., 2002; Pruden et al., 2003). Bonin et al. (2002) followed denitrifying bacteria in a squalene enrichment. The composition of the enrichment culture changed significantly through 8 months of incubation. The isolated strains were compared to the dominant phylotypes on DGGE gels. Eight isolates were obtained from the 12 phylotypes dominating the enrichment cultures. In another example, Burns et al. (2001) isolated *tert*-butyl ether-degrading bacteria and used DGGE to show that the isolates were dominant in compost biofilter enrichments.

2.4.5. Characterization of petroleum-degrading consortia

In some cases, biodegradation of petroleum compounds is performed by a consortium rather than by an individual microorganism. In this case, DGGE can be used to characterize the composition of this consortium without the need to separately cultivate its members, which can be difficult. For example, Koizumi et al. (2002) have used DGGE to study the composition of a mesophilic toulene-degrading consortium and an ethylbenzene-degrading consortium. Their findings revealed that one member of both consortia was affiliated with the family *Desulfobacteriaceae* while the other was related to an uncultured non-sulfate reducing soil bacterium. Similarly, the structure of an anaerobic pentachlorophenol-degrading consortium was revealed by DGGE in a study by Tartakovsky et al. (2001). The consortium was found to consist of *Clostridium* and *Syntrophobacter*/*Syntrophomonas* spp.

2.4.6. Estimation of active members of the community by RT-DGGE

PCR-DGGE analyses can be performed not only with DNA but also with RNA. DNA-based DGGE provides information on the presence of different bacterial populations while RNA-based DGGE gives an indication of which populations are active. The cellular content of ribosomal RNA is related to the recent activity of cells. The first step in RNA-based DGGE is reverse transcription (RT) in order to form double stranded cDNA from single-stranded RNA. PCR is then used to amplify the DNA. This technique has been implemented in several studies to follow changes in active populations upon addition of various organics (Lucas and Hollibaugh, 2001; Schafer et al., 2001). Instead of targeting 16S rRNA molecule, RT-DGGE can be performed, in a similar fashion, using functional genes. For example, in a study by Hoostal et al. (2002), DGGE was performed on RT-PCR amplified fragments of *bph*A1 gene that encodes the large subunit of biphenyl dioxygenase.

2.4.7. Other applications

In general, DGGE has several other applications in addition to the above mentioned ones. It can be used to follow seasonal changes of the microbial communities as well as their spatial and temporal distribution. DGGE has been used to detect the microheterogenity in rRNA encoding genes. Other applications include screening of clone libraries, determining PCR and cloning biases, and comparison of different DNA extraction protocols. More details in the application of DGGE in microbial ecology can be found in Muyzer and Smalla (1998).

2.5. Pitfalls of DGGE

DGGE is like other methods, not free from limitations and biases. Biases can be introduced by sample handling, uneven cell lysis during DNA extraction as well as preferential amplification in the PCR step. The DGGE pattern fails to recover all bacteria but rather give a picture of the dominant species of the community. The obtained sequences from DGGE are short and sometimes difficult to get. Other limitations include the detection of heteroduplex molecules and the co-migration of fragments with different sequences (Muyzer and Smalla, 1998).

3. Microsensor techniques

Contaminated sediments are often characterized by limited diffusion of dissolved substances and high microbial activities. Together they create steep physical and chemical gradients in the contaminated matrix. Such processes can only be investigated with fine-scale analytical techniques. Microsensors, also called microelectrodes, have proven to be very useful tools for this purpose. They allow the simultaneous measurement of several chemical parameters with high spatial resolution and negligible disturbance of the samples. Electrochemical microsensors are characterized by the conversion of a chemical into an electrical signal and can be divided into simple Ag/Ag + half cells, ion-exchanger based electrodes, simple anodes or cathodes with or without polarization and Clark-type gas sensors with ion-impermeable membranes (Kühl and Revsbech, 2001). In microbial ecology, microsensors have been applied to various systems to investigate, e.g. the relationship between photosynthesis and respiration in corals (Al-Horani et al., 2003a) and sediments (Revsbech et al., 1986); internal carbon cycling in corals (Al-Horani et al., 2003b); diffusive boundary layers (Jørgensen and Revsbech, 1985); metabolic processes in microbial mats (Revsbech et al., 1983); and sulfate reduction and sulfide oxidation in trickling-filter biofilms (Kühl and Jørgensen, 1992). Here, we review some applications of microelectrodes in studies related to biodegradation of organic pollutants by cyanobacterial mats.

3.1. Types of microsensors

Microsensors to measure O_2, CO_2, CO_3^{2-}, NO_3^-, NO_2^-, NH_4^+, Ca^{2+}, pH, H_2S and CH_4 have been developed. In this review we will describe only the O_2, H_2S and pH microsensors because of their particular relevance to biodegradation studies.

3.1.1. O_2 microelectrodes

Oxygen is an important parameter in oil biodegradation, as most of the degradation processes occur under oxic conditions. Monitoring oxygen concentration in contaminated sediments is one way to follow the activity of oil-degrading microbial communities, in conjunction with measurements of contaminants concentration. Since oxygen gradients in sediments can be steep, O_2 microelectrodes can be used to measure at the microscale level.

These sensors are similar to the well-known Clark-type oxygen sensors for general laboratory applications, only converted to the μm scale (Revsbech and Ward, 1983) (Fig. 4A–D). They consist of a gold-coated cathode situated behind a silicon membrane and immersed in an electrolyte solution (Fig. 4A). The measuring principle is based on the reduction of O_2 at the cathode. With tip sizes of 1–100 μm and a detection limit of 0.1 μM, such sensors are very stable, insensitive to calcium and magnesium ions and pH independent. The lifetime of the sensors can be longer than a year. Background signals are reduced by a guard cathode, which prevents diffusion of O_2 from behind to the measuring cathode (Revsbech, 1989).

3.1.1.1. Photosynthesis measurements and rate calculations

Simultaneous determination of oxygen concentration and volumetric gross oxygen production rate (i.e. photosynthetic activity) are possible with fast O_2 microsensors (90% response time $t_{90} < 0.5$ s) and the light–dark-shift method (Revsbech and Jørgensen, 1983). The method is based on the determination of the gross oxygen production rate by the decrease in O_2 concentration during a 1 s dark period.

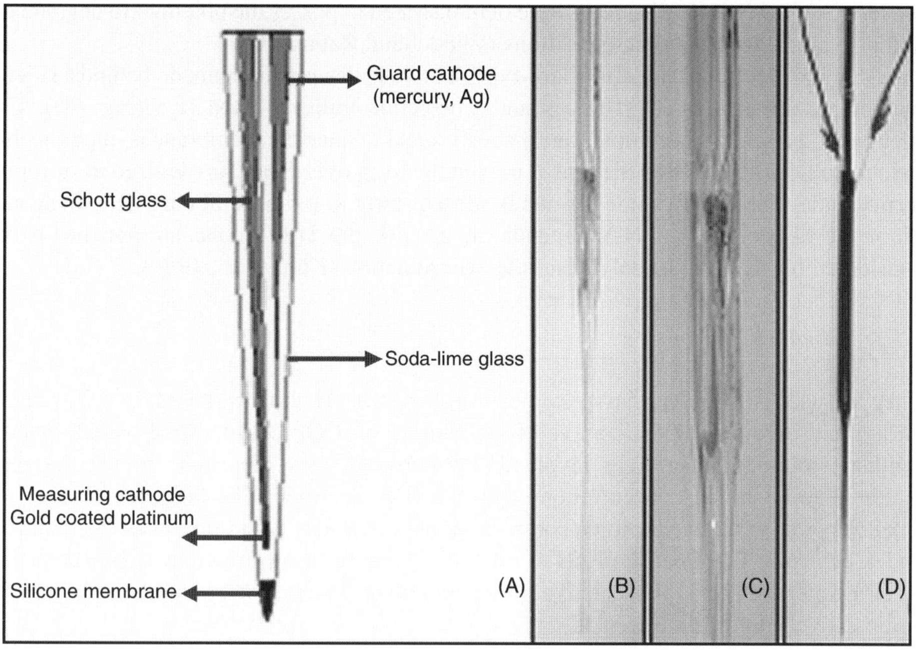

Figure 4. (A) Schematic drawing showing the detailed structure of a Clark-type O_2 microsensor, (B and C) photographs of oxygen microsensor and (D) photograph of H_2S microsensor (photos: Armin Gieseke).

In addition to raw volumetric oxygen concentration profiles (e.g. quantity/m^3), one can calculate production and/or consumption rates for the entire sediment in areal units (e.g. quantity/m^2). These calculations are based on Fick's first law of molecular diffusion and have yielded insights into turnover rates in different sediment zones, e.g. the photic and aphotic zone (Jensen and Revsbech, 1989; Kühl et al., 1996). Such calculations can be done by determination of the first derivative of the concentration values only at the steepest gradients, by integration of the second derivative of the whole profiles or by use of a statistical program (Berg et al., 1998). Subsequently, the different processes contributing to the oxygen equilibrium inside a sample can be calculated based on the mass balance:

Net oxygen production (photosynthesis)
= gross oxygen production (respiration) − oxygen consumption

The effects of oil pollution on photosynthetic and respiratory activity in different layers of contaminated biofilms can be determined by these methods.

3.1.2. H_2S microelectrodes

These microsensors can play an important role in the investigation of deeper layers of contaminated sediments, especially in marine environments. Anaerobic sulfate reduction is a key process in such regions and the evolving sulfide can be measured with such sensors. Sulfate-reducing bacteria were demonstrated to posses the potential to degrade oil components under anaerobic conditions (Widdel and Rabus, 2001).

H_2S microelectrodes are also Clark-type sensors with a platinum anode behind a silicon membrane, tip sizes of 10–100 μm and a detection limit of 1 μM H_2S (Fig. 4D). The lifetime of the sensors is comparatively short (weeks). Since the membrane is impermeable to ions, sulfide diffuses into the sensor only in the form of H_2S, and is oxidized to sulfur by ferricyanide. The ferrocyanide formed is subsequently re-oxidized at the platinum anode. Since the dissociation of H_2S depends on the pH, the H_3O^+ concentration has to be considered for the total dissolved sulfide determination (Kühl et al., 1998).

3.1.3. pH microelectrodes

They are used in biodegradation studies to follow the pH shift triggered by CO_2 release during the aerobic mineralization of pollutants, or CO_2 consumption where photosynthetic microorganisms are involved. Furthermore, they contribute to the determination of total sulfide concentrations with the H_2S electrode. The pH microsensors are small versions of the well-known commercial pH electrodes with tip sizes of 20–200 μm and a detection limit between pH 1 and 14. The measuring principle is based on the electrical potential at a pH-sensitive glass membrane (Hinke, 1969) and the sensor has a comparatively long lifetime.

3.2. Application of microsensors in biodegradation studies

Although microsensors are a powerful tool in microbial ecology, their use in biodegradation studies has so far been limited to studies on cyanobacterial mats and their biodegradation capacities. For example, the influence of carbon source additions on intact cyanobacterial mats was investigated with microsensors for pH, sulfide, oxygen and photosynthesis (Grötzschel et al., 2002a) (Fig. 3, right). Acetate, glycolate and glucose were used as model compounds in order to investigate their effects on the internal carbon cycle in microbial mats. Acetate and glycolate are indigenous substances in those cycles and were degraded in those regions of the mats where they are usually formed. In contrast, the addition of glucose had extreme inhibitory effects on respiration and photosynthesis, probably due to the produced metabolites. The community structure, followed by DGGE, did not exhibit any significant changes upon incubation with these compounds. The responses observed were apparently due to physiological changes in the existing community rather than changes in community composition.

Direct work with oil compounds was also performed. In a laboratory experiment, the degradation of four petroleum model compounds (pristane, *n*-octadecane, phenanthrene and dibenzothiophene) by cyanobacterial mats was investigated (Grötzschel et al., 2002b) (Fig. 3, left). During a period of 4 months the impact of contaminants degradation on oxygen concentration, photosynthetic activity, pH and sulfide concentration was observed with appropriate microsensors. Although the model compounds were degraded at reasonable rates after 18 weeks, there were no significant changes in the physiology of the mats. The degradation happened in the background of the usual high internal turnover rates.

The same microsensors were also used in order to study the impact of oil pollution on the structure and function of a microbial mat (Abed et al., in preparation). The measurements showed a clear decrease in photosynthesis and respiration when oil was added, indicating a toxic effect of oil on the activity of microbial populations. The extent of the decrease in respiration was stronger than in photosynthesis.

In another study microbial mats were grown *de novo* in glass aquaria using an inoculum from Wadi Gaza. The mature mats were subsequently tested for their degradation capacity (Grötzschel et al., in preparation). The previously used model compounds representing different chemical subclasses in petroleum mixtures, namely pristane, phenanthrene, *n*-octadecane and dibenzothiophene, were degraded by the *de novo* established mats. The oxygen demand increased during that period, but the gross oxygen production (photosynthesis) remained constant as measured with microsensors. It was concluded that the aerobic heterotrophic community in those mats played an important role in degradation.

Besides oil and oil compounds, even the removal of other substances like the pesticide 2,4-Dichlorophenoxyacetic acid (2,4-D) by microbial mats was investigated with microsensor techniques (Grötzschel et al., accepted for publication). This short-term investigation revealed the partial inhibition of the phototrophic community by the 2,4-D addition as well as temporary negative effects on sulfate reduction and respiration in the photic zone of the mat.

3.3. A case study

An example of the application of oxygen, sulfide and pH microsensors is given in Figure 5. An easily degradable substance was chosen as a model contaminant to investigate the effect of the degradation process on the physiology of a microbial mat. This experiment was similar to the study described above (Grötzschel et al., 2002a) but with a different sample. Five millimolar of glucose were added to the water phase of a submersed microbial mat. Panel A shows microsensor profiles of the three parameters after contamination. The situation was characterized by strong O_2 demand for aerobic degradation. The oxygen from the overlying water phase was completely consumed within only 400 μm. In the deeper layers, sulfide was produced in large amounts in course of anaerobic degradation and sulfate reduction. These two processes resulted in the pH profile where the low pH was caused by the produced sulfide and CO_2. Panel B shows the situation after removal of the contaminant. The oxygen equilibrium shifted from O_2 consumption to O_2 production. The local maximum at 400 μm below the surface was due to the activity of the photosynthetic autotrophic community, which leads to oxygen oversaturation in the mat and was a characteristic phenomenon in this type of environment. The sulfide concentration strongly decreased to normal levels because of the oxidation of sulfide to sulfur and sulfate by purple sulfur bacteria, colourless sulfur bacteria and chemical oxidation in the new oxic layers of the mats. The pH increased because of the sulfide removal; the consumption of CO_2 for photosynthesis in the upper part of the mat led to a small local maximum at 400 μm depth.

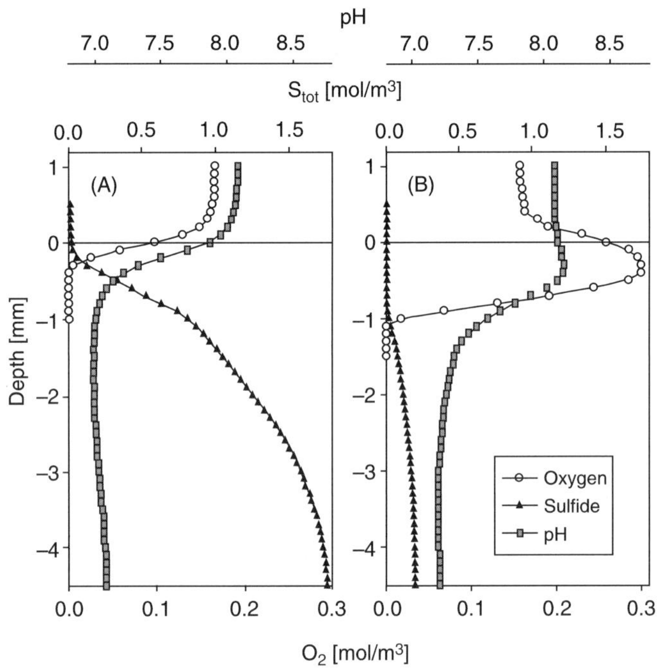

Figure 5. Profiles of pH, oxygen concentration and total sulfide concentration in a microbial mat determined with microelectrodes. (A) After addition of 5 mM glucose; (B) after addition and subsequent exchange of the water phase above the mats.

4. Conclusions

The use of microsensors in the context of biodegradation studies provides information on the role of different functional groups of microorganisms involved in pollutant degradation as well as changes in their physiology. On the other hand, DGGE helps in the identification of key microorganisms responsible for the degradation activities and in monitoring changes in the structure of microbial communities. The major advantage of the two techniques is their possible application directly on field samples, thus revealing the structure and function of microbial populations that operate in the field. Therefore, a combination of these methods represents a powerful alliance of techniques for future investigations in the field of bioremediation.

References

Abed, R.M.M., Safi, N.M.D., Köster, J., de Beer, D., El-Nahhal, Y., Rullkötter, J., Garcia-Pichel, F., 2002. Microbial diversity of a heavily polluted microbial mat and its community changes following degradation of petroleum compounds. Appl. Environ. Microbiol. 68, 1674–1683.

Al-Horani, F.A., Al-Moghrabi, S.M., de Beer, D., 2003a. The mechanism of calcification and its relation to photosynthesis and respiration in the scleractinian coral Galaxea fascicularis. Mar. Biol. 142 (3), 419–426.

Al-Horani, F.A., Al-Moghrabi, S.M., de Beer, D., 2003b. Microsensor study of photosynthesis and calcification in the scleractinian coral, *Galaxea fascicularis*: active internal carbon cycle. J. Exp. Mar. Biol. Ecol. 288 (1), 1–15.

Berg, P., Risgaard-Petersen, N., Rysgaard, S., 1998. Interpretation of measured concentration profiles in sediment pore water. Limnol. Oceonogr. 43 (7), 1500–1510.

Bonin, P.C., Michotey, V.D., Mouzdahir, A., Rontani, J.F., 2002. Anaerobic biodegradation of squalene: using DGGE to monitor the isolation of denitrifying bacteria taken from enrichment cultures. FEMS Microbiol. Ecol. 42 (1), 37–49.

Burns, M.A., Hanson, J.R., Mefford, J., Scow, K.M., 2001. Isolate PM1 populations are dominant and novel methyl *tert*-butyl ether-degrading bacterial in compost biofilter enrichments. Environ. Microbiol. 3, 220–225.

Casamayor, E.O., Schäfer, H., Bañeras, L., Pedrŏs-Aliŏ, C., Muyzer, G., 2000. Identification of and spatio-temporal differences between microbial assemblages from two neighboring sulfurous lakes: comparison by microscopy and denaturant gradient gel electrophoresis. Appl. Environ. Microbiol. 66, 499–508.

Duarte, G.F., Rosado, A.S., Seldin, L., Araujo, W.D., Elsas, J.D.V., 2001. Analysis of bacterial community structure in sulfurous-oil-containing soils and detection of species carrying dibenzothiophene desulfurization (*dsz*) genes. Appl. Environ. Microbiol. 67, 1052–1062.

Friedrich, M., Grosser, R.J., Kern, E.A., Inskeep, W.P., Ward, D.M., 2000. Effect of model sorptive phases on phenanthrene biodegradation: molecular analysis of enrichments and isolates suggests selection based on bioavailability. Appl. Environ. Microbiol. 66 (7), 2703–2710.

Frontera-Suau, R., Bost, F.D., McDonald, T.J., Morris, P.J., 2002. Aerobic biodegradation of hopanes and other biomarkers by crude oil-degrading enrichment cultures. Environ. Sci. Technol. 36 (21), 4585–4592.

Golubic, S., 1992. Microbial mats of Abu Dhabi. In: Margulis, L. (Ed.), Environmental Evolution: Effects of the Origin and Evolution of Life on Planet Earth. Cambridge, Massachusetts, pp. 103–130.

Grötzschel, S., Köster, J., de Beer, D., 2004. Degradation of 2,4-Dichlorophenoxy acetic acid (2,4-D) by a hypersaline microbial mat and related functional changes in the mat community. Microb. Ecol., 48 (2), 254–262.

Grötzschel, S., Abed, R.M.M., de Beer, D., 2002a. Metabolic shifts in hypersaline microbial mats upon addition of organic substrates. Environ. Microbiol. 4 (11), 683–695.

Grötzschel, S., Köster, J., Abed, R.M.M., de Beer, D., 2002b. Degradation of petroleum model compounds immobilized on clay by a hypersaline microbial mat. Biodegradation 13 (4), 273–283.

Head, I.M., Swannell, R.P.J., 1999. Bioremediation of petroleum hydrocarbon contaminants in marine habitats. Curr. Opin. Biotechnol. 10, 234–239.

Hinke, J.A., 1969. Glass microelectrodes in the study of binding and compartmentalization of intracellular ions. In: Lavallee, M.L. (Ed.), Glass Microelectrodes. Wiley, New York, pp. 349–375.

Hoffmann, L., 1996. Recolonization of the intertidal flats by microbial mats after the Gulf war oil spill. In: Krupp, F., Abuzinada, A.H., Nader, I.A. (Eds), A Marine Wildlife Sanctuary for the Arabian Gulf: Environmental Research and Conservation Following the 1991 Gulf War Oil Spill. NCWCD, Riyad and Senckenberg Research Institute, Frankfurt, pp. 96–115.

Höpner, T., Yousef, M., Berthe-Corti, L., Felzmann, H., Struck, H., Al-Thukair, A., 1996. Cyanobacterial mats on oil-polluted sediments – start of a promising self-remediation process? In: Krupp, F., Abuzinada, A.H., Nader, I.A. (Eds), A Marine Wildlife Sanctuary for the Arabian Gulf: Environmental Research and Conservation Following the 1991 Gulf War Oil Spill. NCWCD, Riyad and Senckenberg Research Institute, Frankfurt, pp. 85–95.

Hoostal, M.J., Bullerjahn, G.S., McKay, R.M.L., 2002. Molecular assessment of the potential for in situ bioremediation of PCBs from aquatic sediments. Hydrobiologia. 469 (1–3), 59–65.

Jensen, J., Revsbech, N.P., 1989. Photosynthesis and respiration of a diatom biofilm cultured in a new gradient growth chamber. FEMS Microbiol. Ecol. 62 (1), 29–38.

Jones, W.R., 1998. Practical applications of marine bioremediation. Curr. Opin. Biotechnol. 9, 300–304.

Jørgensen, B.B., Revsbech, N.P., 1985. Diffusive boundary layers and the oxygen uptake of sediments and detritus. Limnol. Oceonogr. 30, 111–122.

Juck, D., Charles, T., Whyte, L.G., Greer, C.W., 2000. Polyphasic microbial community analysis of petroleum hydrocarbon-contaminated soils from two northern Canadian communities. FEMS Microbiol. Ecol. 33 (3), 241–249.

Kasai, Y., Kishira, H., Syutsubo, K., Harayama, S., 2001. Molecular detection of marine bacterial populations on beaches contaminated by the Nakhodka tanker oil-spill accident. Environ. Microbiol. 3 (4), 246–255.

Kleikemper, J., Schroth, M.H., Sigler, W.V., Schmucki, M., Bernasconi, S.M., Zeyer, J., 2002. Activity and diversity of sulfate-reducing bacteria in a petroleum hydrocarbon-contaminated aquifer. Appl. Environ. Microbiol. 68 (4), 1516–1523.

Koizumi, Y., Kelly, J.J., Nakagawa, T., Urakawa, H., El-Fantroussi, S., Al-Muzaini, S., Fukui, M., Urushigawa, Y., Stahl, D.A., 2002. Parallel characterization of anaerobic toluene- and ethylbenzene-degrading microbial consortia by PCR-denaturant gradient gel electrophoresis, RNA–DNA membrane hybridization, and DNA microarray technology. Appl. Environ. Microbiol. 68 (7), 3215–3225.

Kühl, M., Jørgensen, B.B., 1992. Microsensor measurements of sulfate reduction and sulfide oxidation in compact microbial communities of aerobic biofilms. Appl. Environ. Microbiol. 58 (4), 1164–1174.

Kühl, M., Revsbech, N.P., 2001. Biogeochemical microsensors for boundary layer studies. In: Jørgensen, B.B., Boudreau, B.P. (Eds), The Benthic Boundary Layer – Transport Processes and Biogeochemistry. Oxford University Press, New York, pp. 180–210.

Kühl, M., Glud, R.N., Ploug, H., Ramsing, N.B., 1996. Microenvironmental control of photosynthesis and photosynthesis-coupled respiration in an epilithic cyanobacterial biofilm. J. Phycol. 32, 799–812.

Kühl, M., Steuckart, C., Eickert, G., Jeroschewski, P., 1998. A H_2S microsensor for profiling biofilms and sediments: application in an acid lake sediment. Aquat. Microb. Ecol. 15, 201–209.

Lucas, F.S., Hollibaugh., 2001. Response of sediment bacterial assemblages to selenate and acetate amendments. Environ. Sci. Technol. 35 (3), 528–534.

Macnaughton, S.J., Stephen, J.R., Venosa, A.D., Davis, G.A., Chang, Y.-J., White, D.C., 1999. Microbial population changes during bioremediation of an experimental oil spill. Appl. Environ. Microbiol. 65, 3566–3574.

Madsen, E.L., 1998. Theoretical and applied aspects of bioremediation: the influence of microbiological processes on organic contaminant compounds in field sites. In Techniques in microbial ecology. Sayler, G. (ed). New York Oxford: Oxford University Press, pp. 354–407.

Muyzer, G., Smalla, K., 1998. Application of denaturant gradient gel electrophoresis (DGGE) and temperature gradient gel electrophoresis (TGGE) in microbial ecology. Antonie van Leeuwenhoek 73, 127–141.

Muyzer, G., De Waal, E.C., Uitterlinden, A.G., 1993. Profiling of complex microbial populations by denaturant gradient gel electrophoresis analysis of polymerase chain reaction-amplified genes coding for 16S rRNA. Appl. Environ. Microbiol. 59, 695–700.

Muyzer, G., Brinkhoff, T., Nübel, U., Santegoeds, C., Schäfer, H., Wawer, C., 1998. Denaturant gradient gel electrophoresis (DGGE) in microbial ecology. In: Akkermans, A.D.L., van Elsas, J.D., de Bruijn, F.J. (Eds), Molecular Microbial Ecology Manual. Kluwer, Dordrecht, pp. 1–27.

Nübel, U., Garcia-Pichel, F., Muyzer, G., 1997. PCR primers to amplify 16SrRNA genes from cyanobacteria. Appl. Environ. Microbiol. 63, 3327–3332.

Ogino, A., Koshikawa, H., Nakahara, T., Uchiyama, H., 2001. Succession of microbial communities during a biostimulation process as evaluated by DGGE and clone library analyses. J. Appl. Microbiol. 91 (4), 625–635.

Pruden, A., Sedran, M., Suidan, M., Venosa, A., 2003. Biodegradation of MTBE and BTEX in an aerobic fluidized bed reactor. Water Sci. Technol. 47 (9), 123–128.

Revsbech, N.P., Jørgensen, B.B., 1983. Photosynthesis of benthic microflora measured with high spatial resolution by the oxygen microprofile method: capabilities and limitations to the method. Limnol. Oceonogr. 28 (4), 749–756.

Revsbech, N.P., Ward, D.M., 1983. Oxygen microelectrode that is insensitive to medium chemical composition: use in an acid microbial mat dominated by *Cyanidium caldarium*. Appl. Environ. Microbiol. 45 (3), 755–759.

Revsbech, N.P., Jørgensen, B.B., Blackburn, T.H., Cohen, Y., 1983. Microelectrode studies of the photosynthesis and O_2, H_2S, and pH profiles of a microbial mat. Limnol. Oceonogr. 28 (6), 1062–1074.

Revsbech, N.P., Madsen, B., Jorgensen, B.B., 1986. Oxygen production and consumption in sediments determined at high spatial resolution by computer resolution of oxygen microelectrode data. Limnol. Oceonogr. 31 (2), 293–304.

Revsbech, N.P., 1989. An oxygen microsensor with a guard cathode. Limnol & Oceanograph 34 (2): 474–478.

Roling, W.F.M., Milner, M.G., Jones, D.M., Lee, K., Daniel, F., Swannell, R.J.P., Head, I.M., 2002. Robust hydrocarbon degradation and dynamics of bacterial communities during nutrient-enhanced oil spill bioremediation. Appl. Environ. Microbiol. 68 (11), 5537–5548.
Samanta, S.K., Singh, O.V., Jain, R.K., 2002. Polycyclic aromatic hydrocarbons: environmental pollution and bioremediation. Trends Biotechnol. 20, 243–248.
Schafer, H., Bernard, L., Courties, C, Lebaron, P., Servais, P., Pukall, R., Stackebrandt, E., Troussellier, M., Guindulain, T., Vives-Rego, J., Muyzer, G., 2001. Microbial community dynamics in Mediterranean nutrient-enriched seawater mesocosms: changes in the genetic diversity of bacterial populations. FEMS Microbiol. Ecol. 34 (3), 243–253.
Sorkhoh, N., Al-Hasan, R., Radwan, S., 1992. Self-cleaning of the Gulf. Nature 359, 109.
Swannell, R.P.J., Lee, K., McDonagh, M., 1996. Field evaluations of marine oil spill bioremediation. Microbiol. Rev. 60, 342–365.
Tartakovsky, B., Manuel, M.F., Beaumier, D., Greer, C.W., Guiot, S.R., 2001. Enhanced selection of anaerobic pentachlorophenol-degrading consortium. Biotechnol. Bioeng. 73 (6), 476–483.
Timmis, K.N., Pieper, D.H., 1999. Bacteria designed for bioremediation. Trends Biotechnol. 17 (5), 201–204.
Van Hannen E.J., van Agterveld, M.P., Gons, H.J., Laanbroek, H.J., 1998. Revealing genetic diversity of eukaryotic microorganisms in aquatic environments by denaturing gradient gel electrophoresis. Journal of Phycology 34 (2): 206–213.
Vogel, T.M., 1996. Bioaugmentation as a soil bioremediation approach. Curr. Opin. Biotechnol. 7, 311–316.
Watanabe, K., Watanabe, K., Kodama, Y., Syutsubo, K., Harayama, S., 2000. Molecular characterization of bacterial populations on petroleum-contamination groundwater discharged crude oil storage cavities. Appl. Environ. Microbiol. 66 (11), 4803–4809.
Widdel, F., Rabus, R., 2001. Anaerobic biodegradation of saturated and aromatic hydrocarbons. Curr. Opin. Biotechnol. 12, 259–276.
Wilson, G.J., Richter, A.P., Suidan, M.T., Venosa, A.D., 2001. Aerobic biodegradation of gasoline oxygenates MTBE and TBA. Water Sci. Technol. 43 (2), 277–284.

Oil Pollution and its Environmental Impact in the Arabian Gulf Region
M. Al-Azab, W. El-Shorbagy and S. Al-Ghais, editors

Chapter 7

Assessment of minimum sediment concentration for OMA formation using a Monte Carlo model

Ali Khelifa[1,*], Paul S. Hill[1] and Kenneth Lee[2]

[1]*Department of Oceanography, Dalhousie University, Halifax, NS, Canada B3H 4J1*
[2]*Centre for Offshore Oil and Gas Environmental Research, Bedford Institute of Oceanography, Fisheries and Oceans Canada, P.O. Box 1006, 1 Challenger Drive, Dartmouth, NS, Canada B2Y 4A2*

Abstract

Aggregation between suspended sediment grains and crude-oil droplets, which forms oil–mineral aggregates (OMA), is a natural process that enhances dispersion of spilled oil in aquatic environments. The objective of this study is to use a recently upgraded numerical model (MCOMA2) to investigate minimum sediment concentration (MSC) for OMA formation. MSC is defined as the concentration of sediment below which concentration of OMA-stabilized droplets is negligible (less than 1%). Formation of OMA is simulated for sediment concentration varied from 0.5 to 1000 mg/l considering Forties Blend (density of 840 kg/m^3) and FOREF oil (density of 965 kg/m^3). Results show that the MSC varies from 0.5 to 200 mg/l, depending on the size ratio (ratio between the diameter of sediment and the diameter of oil droplets) and the oil type. Its minimum values for both light and heavy oils are obtained when the size ratio is close to 0.4 and 0.55, respectively.

1. Introduction

The term "Oil–mineral aggregates" (OMA) was initially proposed by Lee et al. (1998) to describe oil–mineral fine agglomerations. OMA result from heteroaggregation between suspended sediment and crude-oil droplets dispersed in the water column. The simplest shape of these aggregates consists of an oil droplet coated with sediment grains on its surface. However, complex shapes where more than one droplet contributes to the formation of OMA have been observed from laboratory and field sampling (Lee et al., 1998, 2003; Lee and Stoffyn-Egli, 2001; Stoffyn-Egli and Lee, 2002). OMA form readily if suspended sediment and oil droplets coexist in moderately agitated aquatic environments (Owens, 1999; Khelifa et al., 2002).

This natural process of aggregation between oil droplets and suspended sediment has fascinated researchers dealing with oil spill remediation for decades for two reasons (see Khelifa et al., 2005, for a review). First, sediment grains in an OMA play the role of a natural surfactant and enhance dispersion of spilled oil (Muschenheim and Lee, 2002;

---

*Corresponding author at present address: Environmental Technology Centre, Environment Canada, 335 River Road, Ottawa, ON, Canada K1A OH3. *E-mail address:* ali.khelifa@ec.gc.ca (A. Khelifa).

Owens and Lee, 2003 for a review). When droplets are coated with suspended sediment grains, the droplets do not recoalesce and are stabilized in the water column. Second, recent findings showed that presence of sediment particles on the droplets enhances biodegradation of crude oil (Lee et al., 1997; Jézéquel et al., 1999; Weise et al., 1999).

Factors that affect the formation of OMA include oil composition, composition and abundance of fine sediment, and environmental conditions like temperature, salinity and mixing energy (Lee, 2002 for a review). In particular, sediment size and concentration as well as oil type play significant roles in OMA formation (Bragg and Yang, 1995; McCourt and Shier, 1998, 2000; Owens, 1999; Khelifa et al., 2002, 2003a,b, 2004; Omotoso et al., 2002; Ajijolaiya, 2004). Despite their importance, however, the role of these factors is not well understood yet, and modelling of OMA formation is at its early stage (Payne et al., 1989, 2003; Hill et al., 2002; Khelifa et al., 2002, 2003a,b, 2004, 2005).

This study aims to advance our quantitative understanding of the significance of sediment concentration on OMA formation through application of numerical simulation. Specifically, this study focuses on establishment of quantitative information on the minimum sediment concentration (MSC) below which formation of OMA does not occur. A brief summary of previous studies on the topic is discussed below.

1.1. Minimum sediment concentration

The concept of MSC to form stable OMA, or oil-in-water emulsions in general, has been discussed by many researchers. It is generally agreed that below this minimum concentration, OMA will not form (Khelifa et al., 2003b). Experimental results published by Guyomarch et al. (1999) elucidate this issue. Their data show that a minimum clay concentration is necessary to start the OMA process (some of their data are shown in Figs 1 and 2). This observation is in agreement with what Hoffman and Shier (1999) reported regarding OMA formation in North Slope Rivers; total suspended solids concentrations ranging from 1 to 190 mg/l were found to be insufficient to form OMA. Also, Tambe and Sharma (1993) showed that a minimum concentration of solid particles (calcium carbonate) of about 1% (wt) is required for any stable oil-in-water emulsion to be formed.

McCourt and Shier (1998, 2000) performed laboratory studies on eight Alaskan rivers to show how suspended sediments in the water would interact with crude oil. They found that the amount of oil associated with the sediments was roughly proportional to sediment concentration. In some cases, this amount was negligible. From previous numerical simulations using the simplified numerical model MCOMA1 (Monte Carlo model for OMA formation, first version), Khelifa et al. (2003b) proposed the values of 0.03 and 0.009 as the minimum concentration ratios (the ratio between sediment concentration and the reacting oil) for OMA formation with light (850 g/l) and heavy (975 g/l) oils, respectively. However, floc breakage and aggregation due to differential settling were not considered in the simulations.

1.2. Objectives

The main goal of this study is to use an upgraded numerical model, MCOMA2, to predict the MSC for OMA formation. The model is based on the Monte Carlo method.

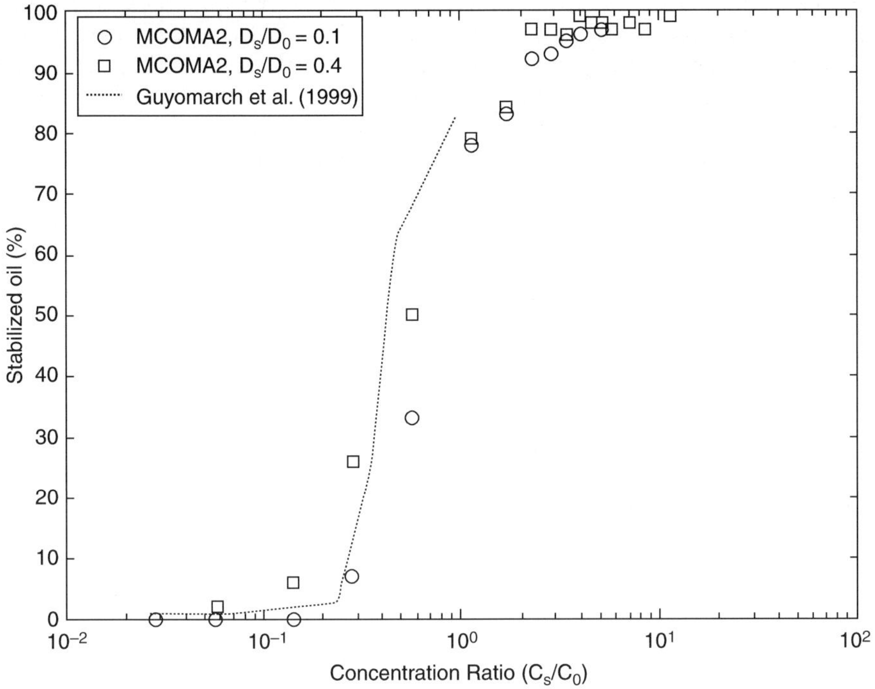

Figure 1. Variations of OMA-stabilized oil concentration (in %) with the concentration ratio and two different size ratios: simulations with Forties Blend crude oil ($\rho_o = 840$ kg/m^3). The dashed line represents laboratory data measured by Guyomarch et al. (1999) using the same oil and Montmorillonite clay with the salinity 35 ppt.

Specifically, the objectives of this study are:

1. To describe the approach used in the model and to corroborate this approach with existing laboratory data published by Guyomarch et al. (1999).
2. To use this model to calculate the MSC for OMA formation for different sediment sizes and two different oils.

2. Monte Carlo method

Several methods have been developed to solve the population balance equation (PBE), which represents the mathematical interpretation of the aggregation/breakage problem (Elimelech et al., 1995 for a review). Our interest is in the Monte Carlo (MC) method because of its well recognized ability to solve multivariate aggregation/breakage problems (Tandon and Rosner, 1999). Discretization problems that result from the direct integration of the PBE in the case of polydispersed population, as occur with the finite difference method (Elimelech et al., 1995), are not an issue in MC simulation. Because of its discrete nature, the MC method adapts itself to growth processes (Lin et al., 2002).

The Monte Carlo (MC) method, which derives its name from extensive use of random numbers, relies on sampling a given statistical ensemble and simulates physical

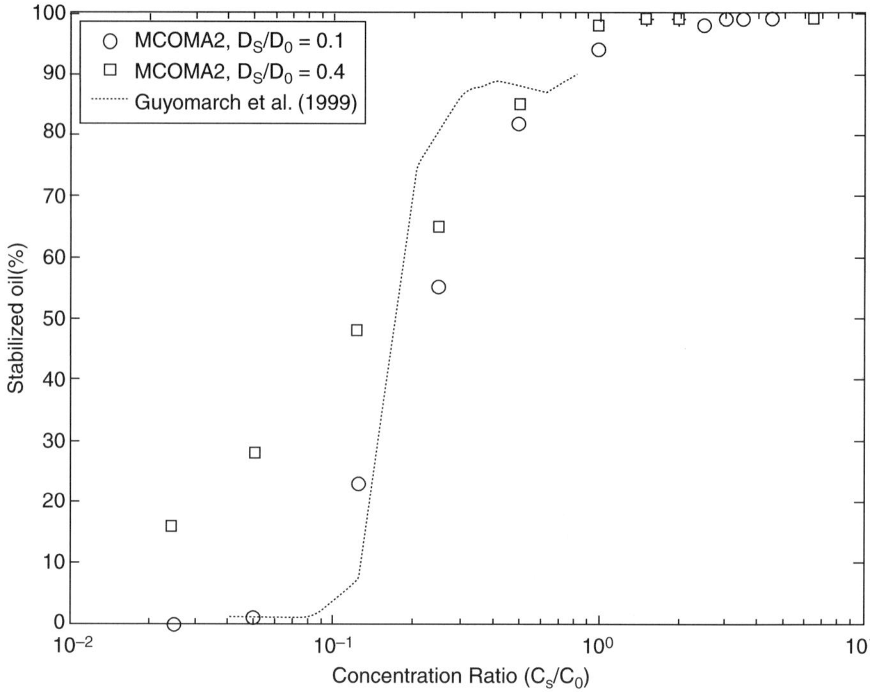

Figure 2. Variations of OMA-stabilized oil concentration (in %) with the concentration ratio and two different size ratios: simulations with FOREF oil ($\rho_o = 965$ kg/m^3). The dashed line represents laboratory data measured by Guyomarch et al. (1999) using the same oil and Montmorillonite clay with the salinity 35 ppt.

processes (e.g. aggregation or breakage) by means of probabilistic tools. For instance, aggregations of two particles to form a third particle, or breakup of a particle to an ensemble of daughter particles are considered as discrete events. At each step of the simulation, a specific event is selected with a probability that is proportional to the rate of its occurrence. The selected event is then applied to one (in case of breakage) or two (in case of aggregation) particles selected stochastically from a predefined array of particles that represents the population. The simulation continues until the equilibrium conditions or the end of the simulation is reached. This method is sometimes referred to as "event driven" MC (Smith and Matsoukas, 1998; Lin et al., 2002). We shall use this method to simulate OMA formation.

3. Description of MCOMA2

The MCOMA2 model uses the constant-volume MC method and the event-driven approach to simulate the formation of OMA. The new model simulates collision between particles due to Brownian diffusion, turbulent motion and differential settling

Essentially, the upgrades introduced into the new model MCOMA2 are: (1) collision rate due to differential settling is integrated into the calculation of the collision frequency between particles. Differential settling takes into account the difference between settling velocities of particles; (2) the formulation previously used in MCOMA1 to calculate the collision rate due to turbulence is replaced by the model proposed by Hill (1992) to take into account the transition from viscous to inertial regimes separated by the Kolmogoroff conditions (Hill, 1992; Hill and Nowell, 1995); (3) the breakage process of sediment flocs is integrated. This process limits the growth of sediment flocs because of the turbulent shear. It is well known that sediment flocs, which result from aggregation between sediment particles, are fragile and their maximum size is limited by various processes including turbulence (Hunt, 1986; Hill, 1998; Hill and McCave, 2001). To integrate floc breakage into the model, floc growth is simulated considering the fractal concept, as proposed by Khelifa and Hill (in press-a). Floc breakage is then simulated by imposing a maximum size for sediment flocs. The maximum size is a function of the turbulent agitation and is introduced at the beginning of each simulation. The density of flocs is tracked using the model proposed by Khelifa and Hill (in press-b).

4. Simulation conditions

A first series of simulations was performed to reproduce laboratory data obtained by Guyomarch et al. (1999) with Forties Blend crude oil (light oil with $\rho_o = 840$ kg/m^3) and FOREF oil (heavy oil with $\rho_o = 965$ kg/m^3). In this series, sediment concentration was varied from 5 to 1000 mg/l for two size ratios of 0.1 and 0.4. Previous studies showed that OMA formation is higher when the size ratio varies between 0.1 and 0.4 (Khelifa et al., 2003b, 2004; Ajijolaiya, 2004). Further simulations were performed to calculate MSC for OMA formation. The size ratio was varied from 0.05 to 4. Oil droplets were monosized with diameter of 10 μm. Sediment concentration was varied from 0.5 to 200 mg/l. All simulations were performed with a constant oil volume concentration of 0.0002. The sediment population was considered as monosized. Simulations were restricted to conditions where the total number of particles was less than about 2.1×10^5, because of the limitation of computational capacity. All simulations were run on a PC (Pentium IV with 1.0 Gb of ram). The simulations were run until all sediment particles were aggregated, or all droplets were stabilized. As discussed previously (Khelifa et al., 2003b), an oil droplet is considered stable if its density becomes larger than or equal to the density of the sea water of 1020 kg/m^3. All the simulated results shown hereafter were obtained from one simulation run.

Typical coastal conditions were simulated. Accordingly, turbulent kinetic energy dissipation rate was set to 0.05 m^2/s^3 (Delvigne and Sweeney, 1988). The corresponding maximum size of sediment flocs was estimated to 54 μm using the model of Parker et al. (1972). The model relates the maximum floc size to the turbulent kinetic energy dissipation rate. The absolute temperature was set to 293 K and the Boltzmann constant was 1.38×10^{-23} J/K.

In all simulations, possible aggregations were sediment–sediment, sediment–droplet and sediment–OMA. Aggregations such as droplet–droplet, droplet–OMA or OMA–OMA were not allowed. This is realistic because under the condition of equilibrium between coalescence and breakage mechanisms, the growth of the droplet size is restricted. Also, we consider that the aggregation efficiency in droplet–OMA and OMA–OMA aggregates is weak. These considerations are supported by the laboratory observations reported by Stoffyn-Egli and Lee (2002).

5. Results and discussion

Variations of OMA-stabilized oil concentration with sediment concentration are shown in Figures 1 and 2 for Forties Blend crude oil and FOREF oil, respectively. Simulation results are shown for size ratios of 0.1 (circle symbol) and 0.4 (square symbol). Laboratory data from Guyomarch et al. (1999) obtained with the same oils and Montmorillonite clay at the salinity of 35 ppt are shown with the dashed lines in the figures. Both Figures 1 and 2 show good agreement between model results and the data. Essentially, the simulations (excluding the results obtained with the size ratio of 0.4 shown in Fig. 2) and the data show that OMA-stabilized oil concentration is very small at low sediment concentrations, increases rapidly with further increase of sediment concentration and then reaches a maximum larger than about 90% at concentration ratios larger than 2 for Forties Blend and 1 for FOREF oil.

The observed rapid enhancement from negligible to significant OMA-stabilized oil concentration with slight increases of the concentration ratio (i.e. sediment concentration because oil concentration is constant in the study) in both the simulation results and the data suggests that MSC for OMA formation exist. For sediment concentrations below MSC, formation of OMA does not occur or OMA-stabilized oil concentration is negligible. Theoretically, this is valid because for a given dispersed droplet population characterized by a diameter D_o, there is a minimum concentration of sediment grains of diameter D_s below which no droplet can be stabilized. Assuming that the stabilizer mechanism is controlled by the settling process (OMA density increases above the seawater density ρ_{sw}), this sediment concentration expressed in a number of sediment grains is simply $(1/D_r)^3(\rho_{sw} - \rho_o)/(\rho_s - \rho_{sw})$, where $D_r = D_s/D_o$ represents the size ratio and ρ_o and ρ_s are the densities of oil and sediment phases, respectively. This expression was established assuming coalescence between sediment grains and the droplet. It shows that MSC exists and varies with the size ratio and the densities of the oil and sediment phases. In the real world, fine sediment grains do not necessarily aggregate with one droplet only, but flocculate with each other and with different suspended droplets. This suggests that the MSC would be much higher than the above estimation.

To investigate how MSC varies with the size ratio and the type of oil, 684 simulations were performed considering the two oils discussed above and various sediment sizes and concentrations. Examples of variations of OMA-stabilized oil concentration with the size ratio at different sediment concentrations are shown in Figures 3 and 4 for Forties Blend and FOREF oils, respectively. Results in these figures show that the percentage of stabilized oil varies considerably with the size ratio. This percentage reaches a maximum when the size ratio is close to 0.55 and 0.4 for Forties Blend and FOREF oils, respectively. Beyond these values of the size ratio, the concentration of stabilized oil decreases rapidly and becomes negligible at size ratios larger than about 3.5. The trend shown by the simulations is well supported by the laboratory data obtained with Flotta oil from UK (Ajijolaiya, 2004; Khelifa et al, 2004). As mentioned earlier, model results obtained with $D_r = 0.4$ show a different trend than those obtained with $D_r = 0.1$ (Fig. 2). This special trend is probably due to the fact that at this size ratio (0.4), OMA formation is maximal for FOREF oil as shown in Figure 4.

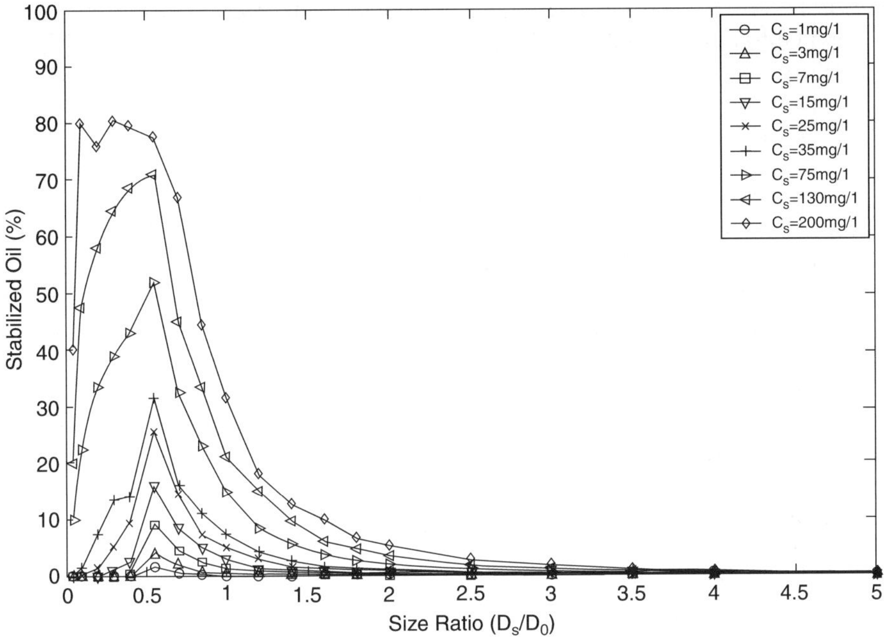

Figure 3. Variations of OMA-stabilized oil concentration (in %) with the size ratio and different sediment concentrations: simulations with Forties Blend crude oil ($\rho_o = 840$ kg/m^3).

From the model results shown in Figures 3 and 4 and others, the MSC is determined assuming the OMA-stabilized oil concentration is less than 0.5% (Fig. 5) or 1% (Fig. 6) of the reacting oil. Values of MSC shown in Figure 6 are slightly higher than those of Figure 5.

Data in both figures showed that the MSC decreases and reaches a minimum value of about 0.5 mg/l at size ratios close to 0.55 and 0.4 for Forties Blend and FOREF oils, respectively.

Beyond these values of the size ratio, the MSC increases rapidly with no apparent effect of the oil type. However, the values of MSC for D_r smaller than about 0.5 are about 3 to 16 times higher for Forties Blend (light oil) than those obtained with FOREF (heavy oil). Overall, results in Figures 5 and 6 show that the MSC varies between about 0.5 and 200 mg/l when D_r varies in the range 0.5 and 4. This range of MSC is in agreement with what Hoffman and Shier (1999) reported regarding OMA formation in North Slope Rivers; total suspended solids concentrations ranging from 1 to 190 mg/l were found not sufficient to form OMA.

From volumetric considerations, it is easy to show that in order to stabilize a droplet with one single sediment grain (the strict MSC), the minimum size ratio should be larger than about 0.35 and 0.52 for Forties Blend and FOREF oils, respectively. These values are very close to those derived from the simulations, i.e. 0.4 and 0.55. As the sediment grain becomes bigger, the mass concentration of sediment increases without further increase in stabilized oil concentration, because one droplet only is still stabilized by the grain. As a result, the MSC increases rapidly with the size ratio as shown in Figures 5 and 6. Similarly, as the grain size becomes smaller than the optimum value, more than one sediment grain is required to stabilize a droplet, and this result in an enhancement of the mass concentration

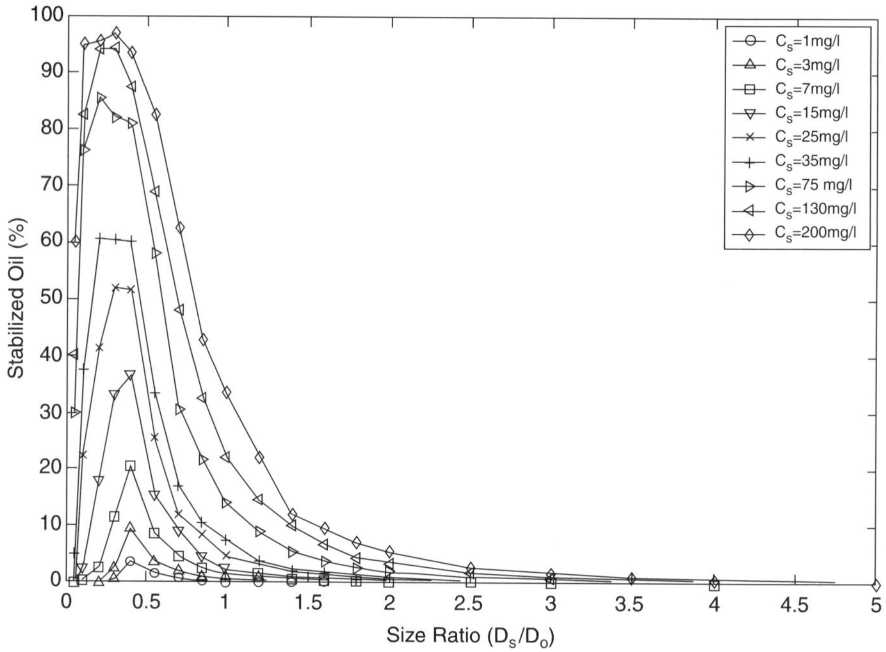

Figure 4. Variations of OMA-stabilized oil concentration (in %) with the size ratio and different sediment concentrations: simulations with FOREF oil ($\rho_o = 965$ kg/m^3).

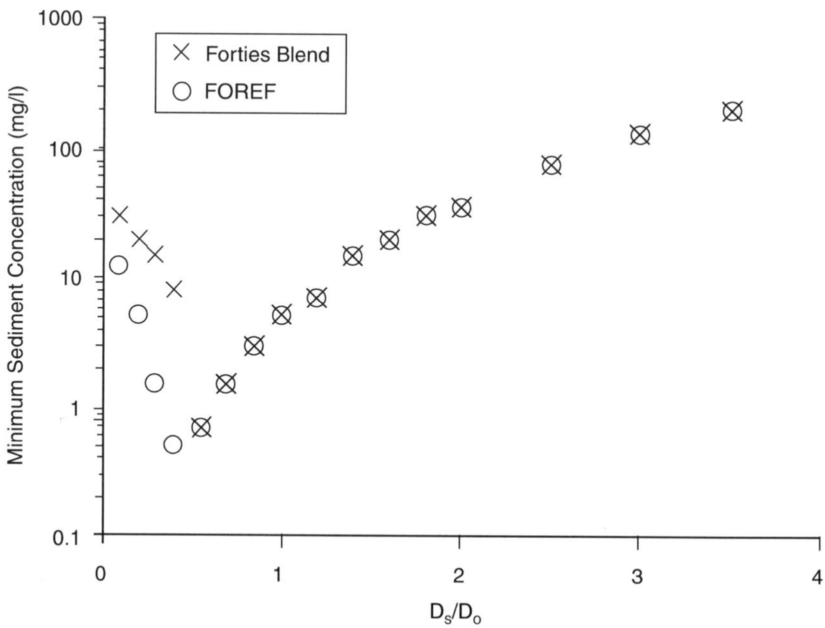

Figure 5. Variations of the minimum sediment concentration with the size ratio: considering 0.5% of the initial oil as the threshold concentration of OMA-stabilized oil.

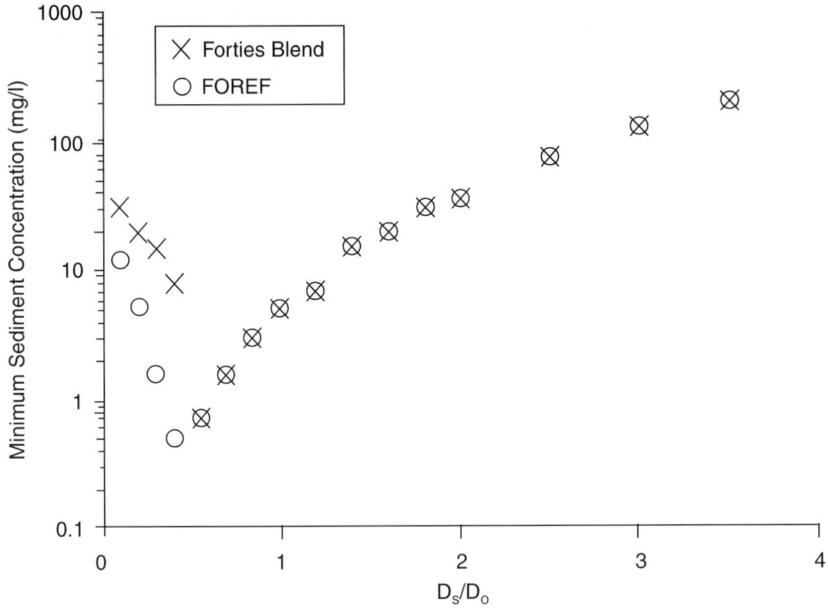

Figure 6. Variations of the minimum sediment concentration with the size ratio: considering 1% of the initial oil as the threshold concentration of OMA-stabilized oil.

of sediment. With further decreases in sediment size, number concentration increases considerably and sediment–sediment aggregation becomes dominant, resulting in the formation of large flocs what can stabilize oil droplets. The mass of sediment grains forming a floc is supposed to increase with number concentration of sediment. Nevertheless, at high number concentrations, surface coverage is expected to play an important role in the stabilization of oil droplets. Previous studies showed that this required a sediment concentration higher than about 1 g/l (Khelifa et al., 2003b).

6. Conclusion

A stochastic model MCOMA2 is used to investigate the effect of sediment size and oil type on the MSC for OMA formation. The study showed that MSC exists below which formation of OMA does not occur or its contribution to stabilization oil droplets is negligible (less than 1% of the reacting oil). Sediment size has strong effects on MSC. The MSC varies between 0.5 and 200 mg/l when the size ratio is between 0.05 and 4. Overall, the curve showing the variations of the MSC with the size ratio is V-shaped. The trough position of the curve, where the MSC reaches its minimum of about 0.5 mg/l, is oil-dependent. It was obtained at the size ratio of about 0.4 for Forties Blend and 0.55 for FOREF oil. When the size ratio is less than about 0.4, oil type affects MSC. At larger size ratios, MSC is not affected by oil type.

The range 0.5–200 mg/l for MSC obtained in this study corresponds to background sediment concentrations in the coastal zones. This is further evidence that OMA form readily in such relatively agitated environments if oil droplets are formed. Comparisons with laboratory data from different sources validated the performance of the MOCMA2 model, yet more laboratory and field data are required to corroborate the results of this study.

In terms of predicting the environmental persistence of oil spilled in coastal environments, while MCOMA2 can be refined further, it is ready for application to calculate concentration of OMA under any conditions if the maximum size allowed for sediment flocs, sediment concentration, mean droplet and sediment sizes are known. Refinements to the model are primarily related to parameterization of the sticking efficiency after collision between sediment particles and oil droplets, modelling of the maximum size of sediment flocs as a function of environmental conditions and sediment properties, and most importantly, modelling of droplet formation for oil stranded on the shore and within slicks prior to their reaction with sediments.

Acknowledgements

This study is part of a program to develop numerical models to simulate heteroaggregation of oil and sediment after an accidental oil spill. It is supported by the Petroleum Research Atlantic Canada (previously Atlantic Canada Petroleum Institute) and the Natural Sciences and Engineering Research Council of Canada.

References

Ajijolaiya, L.O., 2004. The effects of mineral size and concentration on the formation of oil–mineral aggregates. Master Thesis, Dalhousie University, Halifax, Canada, p. 80.

Bragg, J.R., Yang, S.H., 1995. Clay-oil flocculation and its role in natural cleansing in Prince William Sound following the Exxon Valdez oil spill. In: Wells, P.G., Butler, J.N., Hughes, J.S. (Eds), Fate and Effects in Alaskan Waters, ASTM STP 1219. American Society for Testing and Materials, Philadelphia, pp. 178–214.

Delvigne, G.A.L., Sweeney, C.E., 1988. Natural dispersion of oil. Oil Chem. Pollut. 4, 281–310.

Elimelech, M., Gregory, J., Jia, X., Williams, R.A., 1995. Particle Deposition and Aggregation. Butterworth-Heinemann, Massachusetts, p. 441.

Guyomarch, J., Merlin, F., Bernanose, P., 1999. Oil interaction with mineral fines and chemical dispersion: behaviour of the dispersed oil in coastal or estuarine conditions. Environment Canada's 22nd Arctic and Marine Oilspill (AMOP) Technical Seminar, Calgary, Alberta, Canada, pp. 137–149.

Hill, P.S., 1992. Reconciling aggregation theory with observed vertical fluxes following phytoplankton blooms. J. Geophys. Res. 97 (C2), 2295–2308.

Hill, P.S., 1998. Controls on floc size in the sea. Oceanography 11, 13–18.

Hill, P.S., McCave, I.N., 2001. Suspended particle transport in benthic boundary layers. In: Boudreau, B.P., Jorgensen, B.B. (Eds), The Benthic Boundary Layer: Transport Processes and Biogeochemistry. Oxford University Press, Oxford, pp. 78–103.

Hill, P.S., Nowell, A.R.M., 1995. Comparison of two models of aggregation in continental-shelf bottom boundary layers. J. Geophys. Res. 100 (C11), 22749–22763.

Hill, P.S., Khelifa, A., Lee, K., 2002. Time scale for oil droplet stabilization by mineral particles in turbulent suspensions. Spill Sci. Technol. Bull. 8 (1), 73–81.

Hoffman, R.W., Shier, L., 1999. Additional investigations of oil–sediment interactions in Alaskan north slope and TAP area rivers. Environment Canada's 22nd Arctic and Marine Oilspill (AMOP) Technical Seminar, Calgary, Alberta, Canada.

Hunt, J.R., 1986. Particle aggregate breakup by fluid shear. In: Mehta, A.J. (Ed.), Estuarine Cohesive Sediment Dynamics. Springer, Berlin, pp. 85–109.

Jézéquel, R., Merlin, F.-X., Lee, K., 1999. The influence of microorganisms on oil–mineral fine interactions in low energy coastal environment. In: Proceedings of the 1999 International Oil Spill Conference. American Petroleum Institute, Washington, DC, pp. 771–775, Publication no. 4686B.
Khelifa, A., Hill, P.S., in press-a. Kinematic assessment of floc breakage using a Monte Carlo model. J. Hydraul. Res., IAHR.
Khelifa, A., Hill, P.S., in press-b. Models for effective density and settling velocity of flocs. Submitted to J. Hydraul. Res., IAHR.
Khelifa, A., Stoffyn-Egli, P., Hill, P.S., Lee, K., 2002. Characteristics of oil droplets stabilized by mineral particles: the effect of oil types and temperature. Spill Sci. Technol. Bull. 8 (1), 19–30.
Khelifa, A., Stoffyn-Egli, P., Hill, P.S., Lee, K., 2003a. Characteristics of oil droplets stabilized by mineral particles: the effect of salinity. 2003, International OilSpill Conference, Vancouver, BC, Canada, pp. 1–8.
Khelifa, A., Hill, P.S. Lee, K., 2003b. A stochastic model to predict the formation of oil–mineral aggregates. 26th Proceedings of Arctic and Marine OilSpill Program (AMOP) Technical Seminar, Environment Canada, Victoria, BC, Canada 2, 893–909.
Khelifa, A., Hill, P.S., Lee, K., 2004. Modelling the effect of sediment size on OMA formation. 27th Proceedings of Arctic and Marine OilSpill Program (AMOP) Technical Seminar, Environment Canada, Edmonton, Alberta, Canada.
Khelifa, A., Hill, P.S., Lee, K., 2005. The role of oil-sediment aggregation in dispersion and biodegradation of spilled oil. In: Al-Azab, A., El-Shorbagy, W., Al-Ghacs, S. (Eds) Oil Pollution and its Enironmental Impact in the Arabian Gulf Region. Elsevier, London, pp. 131–145.
Lee, K., 2002. Oil-particle interactions in aquatic environments: influence on the transport, fate, effect and remediation of oil spills. Spill Sci. Technol. Bull. 8 (1), 3–8.
Lee, K., Stoffyn-Egli, P., 2001. Characterization of oil–mineral aggregates. In: Proceedings of the 2001 International Oil Spil Conference. American Petroleum Institute, Washington, DC, pp. 991–996.
Lee, K., Lunel, T., Wood, P., Swannell, R., Stoffyn-Egli, P., 1997. Shoreline cleanup by acceleration of clay-oil flocculation processes. In: Proceedings of the 1997 International Oil Spill Conference. American Petroleum Institute, Washington, DC, pp. 235–240, Publication no. 4651.
Lee, K., Stoffyn-Egli, P., Wood, P.A., Lunel, T., 1998. Formation and structure of oil–mineral fines aggregates in coastal environments. Proceedings of the 21st Arctic and Marine Oilspill Program (AMOP) Technical Seminar. Environment Canada, Ottawa, Ontario, pp. 911–921.
Lee, K., Stoffyn-Egli, P., Tremblay, G.H., Owens, E.H., Sergy, G.A., Guenette, C.C., Prince, R.C., 2003. Oil–mineral aggregate formation on oiled beaches: natural attenuation and sediment relocation. Spill Sci. Technol. Bull. 8 (3), 285–296.
Lin, Y., Lee, K., Matsoukas, T., 2002. Solution of the population balance equation using constant-number Monte Carlo. Chem. Eng. Sci. 57, 2241–2252.
McCourt, J., Shier, L., 1998. Interaction between oil and suspended particulate matter in the Yukon river. Environment Canada's 21st Arctic and Marine Oilspill Program (AMOP) Technical Seminar, Ottawa, Ontario, Canada, pp. 77–88.
McCourt, J., Shier, L., 2000. Preliminary findings of solids interaction in eight Alaskan rivers. Environment Canada's 23rd Arctic and Marine Oilspill Program (AMOP) Technical Seminar, Vancouver, BC, Canada.
Muschenheim, D.K., Lee, K., 2002. Removal of oil from the sea surface through particulate interactions: review and prospectus. Spill Sci. Technol. Bull. 8 (1), 9–18.
Omotoso, O.E., Munoz, V.A., Mikula, R.J., 2002. Mechanisms of crude oil–mineral interactions. Spill Sci. Technol. Bull. 8 (1), 45–54.
Owens, E.H., 1999. The interaction of fine particles with stranded oil. Pure Appl. Chem. 71 (1), 83–93.
Owens, E.H., Lee, K., 2003. Interaction of oil and mineral fines on shorelines: review and assessment. Mar. Pollut. Bull. 47 (9-12), 397–405.
Parker, D.S., Kaufman, W.J., Jenkins, D., 1972. Floc breakup in turbulent flocculation processes. J. Sanit. Eng. Div., 79–99, ASCE: 98 (SA1).
Payne, J.R., Clayton, J.R., McNabb, G.D., Jr., Kirstein, C.L., Clary, C.L., Redding, R.T., Evans, J.S., Reimnitz, E., Kempema, E.W., 1989. Oil–ice–sediment interactions during freezeup and breakup. U.S. Department of Commerce, NOAA, OCSEAP. Final Report 64, p. 382.
Payne, J.R., Clayton, J.R. Jr., Kirstein, B.E., 2003. Oil/suspended particulate material interactions and sedimentation. Spill Sci. Technol. Bull. 8 (2), 201–221.

Smith, M., Matsoukas, T., 1998. Constant-number Monte Carlo simulation of population balances. Chem. Eng. Sci. 53 (9), 1777–1786.

Stoffyn-Egli, P., Lee, K., 2002. Formation and characterization of oil–mineral aggregates. Spill Sci. Technol. Bull. 1, 31–44.

Tambe, D.E., Sharma, M.M., 1993. Factors controlling the stability of colloid-stabilized emulsions. I. An experimental investigation. J. Colloid Interface Sci. 157, 244–253.

Tandon, P., Rosner, D.E., 1999. Monte Carlo simulation of particle aggregation and simultaneous restructuring. J. Colloid Interface Sci. 213, 273–286.

Weise, A.M., Nalewajko, C., Lee, K., 1999. Oil-mineral fine interactions facilitate oil biodegradation in seawater. Env. Technol. 20, 811–824.

III

Modeling the fate of pollutants and oil slicks in marine water

Oil Pollution and its Environmental Impact in the Arabian Gulf Region
M. Al-Azab, W. El-Shorbagy and S. Al-Ghais, editors

Chapter 8

A three-dimensional simulation of pollutants transport in the Abu-Qir Bay, East Alexandria, Egypt

E. Al-Hogaraty[1,*], W. Hamza[2], J. Koponen[3], J. Sarkkula[4] and Alam El-Din[5]

[1]*Arab Academy for Science and Technology and Maritime Transport, P.O. Box 52212, Dubai, United Arab Emirates*
[2]*Biology Department, Faculty of Science, United Arab Emirates University, P.O. Box 17551, Al-Ain, United Arab Emirates*
[3]*Environmental Impact Assessment Center of Finland Ltd., Tekniikantie 21 B, FIN-0215 Espoo, Finland*
[4]*Finnish Environment Institute (FEI), P.O. Box 140, FIN-00251 Helsinki, Finland*
[5]*Oceanography Department, Faculty of Science, Alexandria University, Alexandria, Egypt*

Abstract

Abu-Qir Bay is a semicircular basin in the eastern part of Alexandria City, Egypt. The Bay has a maximum depth of 15 m, and surface area of 500 km^2. The bay is suffering from a serious pollution problem caused by the inflow of brackish water from Lake Edku and the poor-quality water discharged from the Rosetta Nile branch. The variations in wind regimes and seasonal fluctuations in the horizontal distributions of water temperature and salinity have largely affected water circulation in the Bay. The present study investigates water circulation in Abu-Qir Bay and its role in pollutants transport, with the use of numerical modeling techniques. Modeled pollutants include toxic metals and hydrocarbons (dissolved and dispersed petroleum hydrocarbons (DDPH)). The flow velocities induced by several combinations of wind and discharge of the Rosetta Nile branch, Lake Edku and Tabia Pump Station (TPS) were calculated at time steps of 30 s for a period of one month per season each year. The observation point (monitoring station) was chosen near the Abu-Qir Harbor. The calibration and sensitivity analysis of the model were carried out with the use of field-measured pollutant concentrations of toxic metals and hydrocarbons (DDPH) and water circulation data for the Abu-Qir Bay.

1. Introduction

Alexandria is the second largest city in Egypt, it is considered as the largest resort on the Egyptian Mediterranean Coast.

---

*Corresponding author. *E-mail address:* eslamocean@yahoo.com (E. Al-Hogaraty).

Due to the rapid increase and development of the growing population activities, the coastal water of Alexandria City is threatened by accumulation of pollutants from terrestrial, industrial and domestic sources. In fact, huge amounts of polluted disposals are discharged into several semi-enclosed basins in Alexandria such as: El-Mex Bay, the Eastern Harbor and Abu-Qir Bay. Therefore, the study of water pollution along the Alexandria coast has a national interest.

Abu-Qir Bay is a shallow semicircular basin, located at about 35 km north-east of Alexandria City between Abu-Qir headland (30°06′E, 31°32′5″N) in the south-west and the Rosetta Mouth of the River Nile (30°35′E, 31°50′N) in the northeast. It has a total area of about 500 km^2 with a shoreline length of about 50 km. The maximum depth recorded is 15 m, which decreasing gradually towards the shore as shown in Figure 1.

The Bay is mostly affected by different sources of land drainage including: industrial, agricultural and domestic disposals from Rosetta Mouth of the River Nile, El-Maadiya outlet of Lake Edku and El-Tabia Pumping Station (TPS) which dump both of organic and

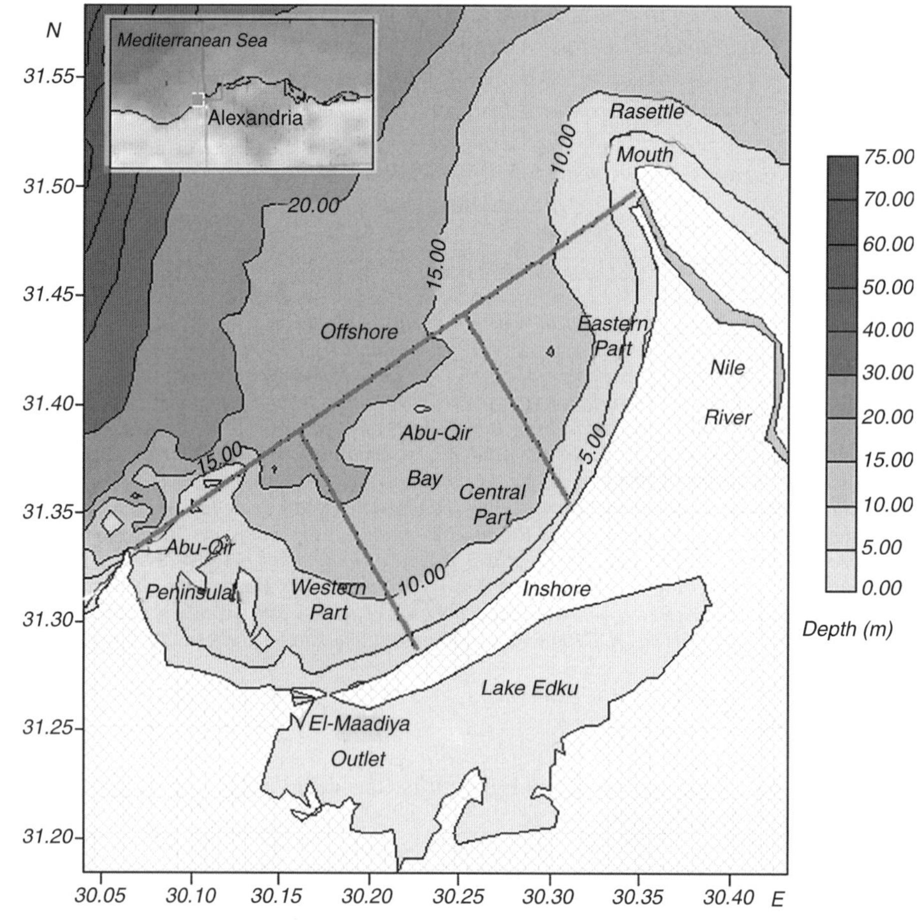

Figure 1. Model domain and Bathymetric Map of Abu-Qir Bay.

inorganic wastes from about 22 different factories representing food processing and canning, paper industry, fertilizer industry, textile manufacturing and petrochemical industry that are surrounding the Bay.

The main objectives of the present investigation is to study the water circulation in Abu-Qir Bay and its effect on pollutants transport especially those derived from land-based sources using numerical modeling technique in order to find the limits to minimize these effects.

2. Initial set-up and installation

The hydrographic survey carried out during a one-year program (1990) and the MED4 seasonal climatology dataset was used as the main dataset to feed the water quality model by the initial fields for temperature, salinity and velocity.

The collected data from the previous investigations dealing with discharge and concentrations of pollutants in Abu-Qir Bay in front of TPS, Bogaz El-Maadiya and Rosetta mouth have been used to determine the load of major pollutants in the Bay.

3. Methodology

The water quality model (EIA) application to simulate Abu-Qir Bay area was subdivided into 57×82 horizontal grid squares using equal mesh resolution of 800 m. In the vertical direction the water column was divided into nine layers with layer interfaces at 1, 3.5, 7.2, 15, 25, 40, 75, 150 and 250 m depths. The changes within each of the grid boxes and the differences and fluxes between them are calculated which is displayed in Figure 2.

Model bathymetry was obtained from the Admiralty Chart, by bilinear interpolation of the depth data into the model. The recorded wind data from January to December 1990 were used to run the model (Meligy, 2000) as shown in Figure 3.

The flow velocities induced by several combinations of wind, Nile River discharge through Rosetta Mouth, Lake Edku and TPS were calculated with a time step of 30 s for periods of about one month for each season of the year. The observation point was set near Abu-Qir Harbor.

4. Study of pollutant transport

In this study, both particulate and dissolved lead (Pb) have been represented as total lead (Pb TOT), where petroleum hydrocarbons have been represented as dissolved and dispersed petroleum hydrocarbons (DDPH). Moreover, the seasonal variations in the total load of the targeted variables were expressed as kg/day.

5. Lead

In Abu-Qir Bay, the particulate form of lead was dominant in both surface and bottom water layers (Younes et al., 1997, 1999). There was an increase in dissolved lead values for both

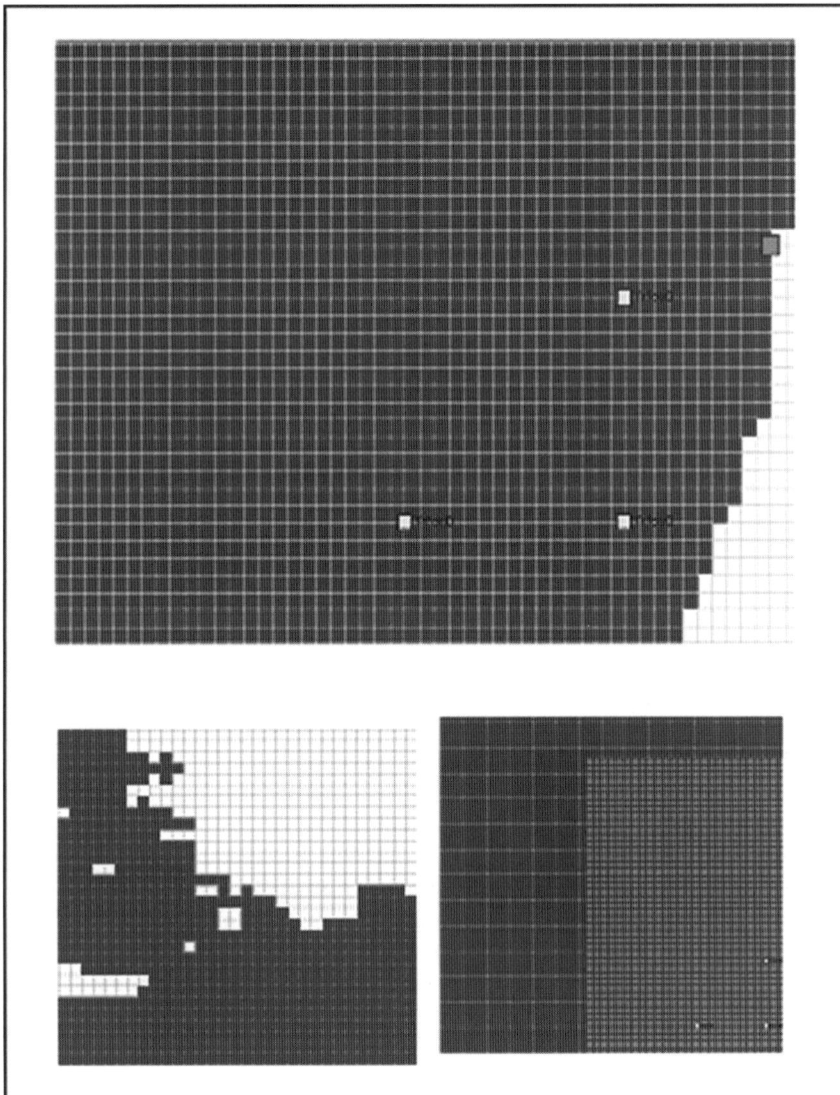

Figure 2. Model Grid Pattern that is used in EIA Model for Abu-Qir Bay.

water layers toward the seaward this indicating that the pumping station was dominantly introducing lead to the inshore locations of the Bay (El-Haridi, 1994; Younes, 1997).

The annual mean of the dissolved lead in the Bay is 6.9 mg/m^3 and most of the total lead was present as particulate lead in the surface and bottom waters of the Bay (Younes, 1997). The particulate lead values showed a decrease in the seaward direction. The annual mean of the particulate lead in the Bay is 14.3 mg/m^3. Where it is mainly released from the bottom sediments and\or from mobilization of the decayed dead organisms.

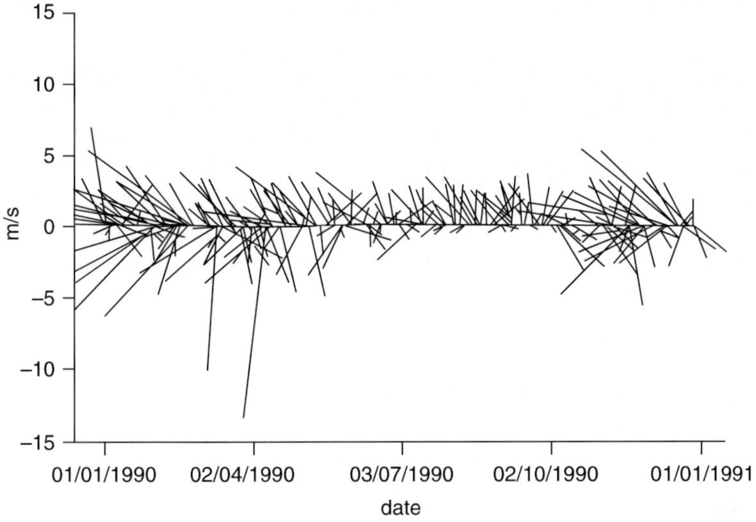

Figure 3. Recorded wind data from January to December 1990.

The amounts of dissolved (DPb) and particulate lead (PPb) discharged through TPS to the Bay water are 7.96 ton year^{-1} and 12.2 ton year^{-1}, respectively (Younes, 1997); however, the annual lead input to the Bay through the discharges through Lake Edku is about 6.94 + 2.45 ton year^{-1}.

6. Dissolved/dispersed petroleum hydrocarbons

Abu-Qir Bay is marked as the most contaminated area along the Alexandria coastline, because of the highly mean concentration of DDPH (2.31 μg/g) in algae, (0.344–0.408 μg/g) inshore sediments and (2.517 μg/l) in water. Oil pollution destroys commercial fisheries, beaches and affects tourism activities, in Abu-Qir Bay.

The general picture of the distribution of biological assemblages in Abu-Qir Bay shows a relatively high diversity in species composition and intensity particularly at the offshore stations. This area is subjected to less eutrophic effect because the industrial wastes diluted through the circulation of seawater and current system in the Bay and exchange with the open seawater.

The lowest value of species diversity was recorded at El-Tabia and Boughaz areas representing the highest eutrophication regions. The productivity in this area showed maximum values, which is nearly more than six times that of offshore stations (Shabara, 2000).

7. Model structure

The Environmental Impact Assessment (EIA) model is a three-dimensional (3D) model of water currents and transport of pollutants including the effluent from different sources. The model was applied to Abu-Qir Bay to discuss the hydrodynamics and pollutants transport in the area. In this model, transport of pollutant is based on wind and river flow-induced currents calculated with the 3D baroclinic multilayer model (Simons, 1980;

Sarkkula and Virtanen, 1983; Koponen, 1991; Sarkkula, 1991; Koponen et al., 1992; Sarkkula et al., 1998).

The water masses were treated as horizontal layers interfaced by levels at selected depths. Horizontally the model area is subdivided into rectangles with arbitrary mesh intervals in both directions. Explicit finite-difference schemes are used for the numerical solution of flow velocities and water level elevations. The generated current in the model are affected by the following factors: wind force, atmospheric pressure at the surface, conservation and incompressibility of water, internal friction (viscosity), transport of velocity differences with water currents (advection), coriolis force, density differences, water level gradients (hydrostatic pressure) and bottom friction.

In other words, these forcing factors used in the model can be summarized as surface shear (spatially and temporally varying winds), river flows, boundary surface elevations (tides) and density effects. Model parameters include bottom drag coefficient, latitude (coriolis), viscosity and diffusion (Kibar, 1997).

In practice the calculation of water currents is detached from that of material transport and water quality (parallel attempts by Virtanen and Koponen, 1985; parallel attempts by Halfon et al., 1990, revaluation in Koponen et al., 1992; Soussi et al., 1995).

Mathematically, the physical interactions and mechanisms affecting water currents are described with Reynold's equations of motion, with the equation of hydrostatic pressure as a proper approximation of its vertical components and with the equation of continuity:

$$\frac{\partial u}{\partial t} = \frac{-\nabla \cdot p}{\rho_0} + \nabla \cdot (K \cdot \nabla) u - 2\Omega u - u \cdot \nabla u + F \tag{1}$$

$$P = P_a + g \cdot (\eta - z) \cdot \rho_0 + g \int_z^{\eta} (\rho + \rho_0) \mathrm{d}z \tag{2}$$

$$\frac{\partial u}{\partial x} + \frac{\partial v}{\partial y} + \frac{\partial w}{\partial z} = 0 \tag{3}$$

where:

Variable	Significant
u	Flow velocity vector
t	Time
η	Surface level elevation
P	Pressure
ρ_0	Average water density
Ω	Angular velocity of the earth rotation
P_a	Air pressure
K	Dispersion matrix of momentum
g	Gravitational acceleration
F	External forcing vector = (0,0, − g)

The flow fields (UL) induced by winds and water discharges Q are calculated in advance from several levels of water surface (η_L). When the water level is changed, the flow fields for transport calculation are changed accordingly. At the same time, surface area covered with water is updated. For any water level (η) between two levels of flow field calculation (η_L) and (η_{L+1}) the flow field U.

i.e. the flow velocities $u = (u, v, w)$ at any location (m, n, k) of 3D-grid will be:

$$U = \frac{\eta - \eta_{L+1}}{\eta_L - \eta_{L+1}} U^L + \frac{\eta_L - \eta}{\eta_L - \eta_{L+1}} U^{L+1} \tag{4}$$

Calculation of water quality: release, transport, mixing resettling, the changes of water quality and bottom properties are coupled with each other in most complex net interactions. In the model, these are gathered within the transport equation, with internal loading as one of its boundary conditions:

$$\frac{\partial c}{\partial t} = -u^S \frac{\partial c}{\partial x} - v^S \frac{\partial c}{\partial y} - w^S \frac{\partial c}{\partial z} + \frac{\partial}{\partial x}\left(D_x \frac{\partial c}{\partial x}\right) + \frac{\partial}{\partial y}\left(D_y \frac{\partial c}{\partial y}\right) + \frac{\partial}{\partial z}\left(D_z \frac{\partial c}{\partial z}\right) + \frac{(\mathrm{qL})}{\partial n} + R(T, c, \ldots) \tag{5}$$

$$\mathrm{qL} = P + R + D + B + A \tag{6}$$

where

Variable	Significant
c	Concentration (or strength) of any quantity computed
u^s, v^s, w^s	Flow velocity components obtained from the equation of 3D-grid currents
D_x, D_y, D_z	Dispersion coefficient components of concentrations
qL	Loading discharge (from point source P, rivers R, drainage area D, bottom B and through water surface A)
n	Discharge direction coordinate
R	Changes from internal (biogeochemical) reactions within water

$$\begin{aligned} &\text{At shore: } (u^S, v^S, w^S)n = 0, (D_x, D_y, D_z) = 0, \\ &\text{At inflows: } \mathrm{qL}R = \mathrm{Q.cin}, \\ &\text{At bottom: } \mathrm{qL}B = \mathrm{E0} + \mathrm{E1} + \mathrm{E2} \end{aligned} \tag{7}$$

where

Variable	Significant
$(x, y, z)n$	Perpendicular component of vector (x, y, z)
cin	Concentration at the inflow Q
Ei	Release rates in internal loading from bottom

In the solution, the simultaneous complex interactions are governed by considering each process after the other with its immediate consequences and going through all of the processes step by step with short intervals until the desired time period as the sum of the time-steps for each process is equally expired (Virtanen et al., 1994).

8. Results and discussion

Abu-Qir Bay lies to the east of Alexandria City and is considered one of the most environmentally sensitive areas along the Alexandria coast. Enormous amount of wastewaters are discharged from land-base sources into the coastal waters, with harmful and toxic effects, leading to serious impacts on the marine environment off Alexandria. This study aims at predicting the distribution of the pollutants that help to understand fate and impact of pollutants on the marine biota and human life.

Generally wind is the predominated parameter that cause a significant effects on the surface water circulation of the Abu-Qir Bay where the influence of water discharge from each Rosetta Mouth of the River Nile and Lake Edku is clearly observed near the outfalls all over the year.

From the results of the numerical simulations it has been found that the targeted variables have different behaviors at the surface water such as at winter, these variables had two privileged flowing directions. The first is parallel to the coast along the north-eastern region of the Bay while the most important direction was flowing south to the western and edge of the central part of the Bay, which seems to be more often impacted as shown in Figure 4a–c (Figures 5–9).

During spring and under the effect of northeast prevailing winds the water circulation plied the targeted variables to be located between TPS and Lake Edku with highly concentration records of lead and DDPH.

Under the influence of the north and north-easterly winds prevailing during summer, most of the targeted variables derived from brackish water, were directed more eastwards. The southwest area of Abu-Qir Bay sustained high concentrations of pollutants (lead and DDPH) due to the maximum discharge from TPS during summer. The traces of targeted variables are normally peaked up during summer time following the peak of the maximum sewage discharge. This finding agreed with the other chemical observations along the Bay, especially that of Abdel-Moati (2001).

During autumn, under the influence of the north-easterly wind and the high flow from TPS, surface water is drifted westwards to cover most of the south-western side of the Bay. The high concentrations recorded for lead, proves that this area (in front of the lake–sea connection) became under stress as long as a considerable amount of industrial wastewater is discharged into the water body causing harmful changes in its water quality.

Also, the results of the simulation represent a better and necessary guide to help the decision makers and specialists on the environmental field towards a better management and protection of the environment of the Bay, by apply the following:

1. A step-wise plan (at least 10% reduction in the present discharge per year) is essential for controlling the load of lead (Pb), total suspended matter (TSM), oil and other pollutants from TPS.
2. Establishment of a simple long-term strategy for monitoring Abu-Qir Bay that is based on an individual monitoring station off TPS and develop a database for the available information on water quality from land based sources.
3. Continuous measurements of various environmental and hydrodynamic parameters such as currents, waves, sea level changes, sediments transportation and coastal zone

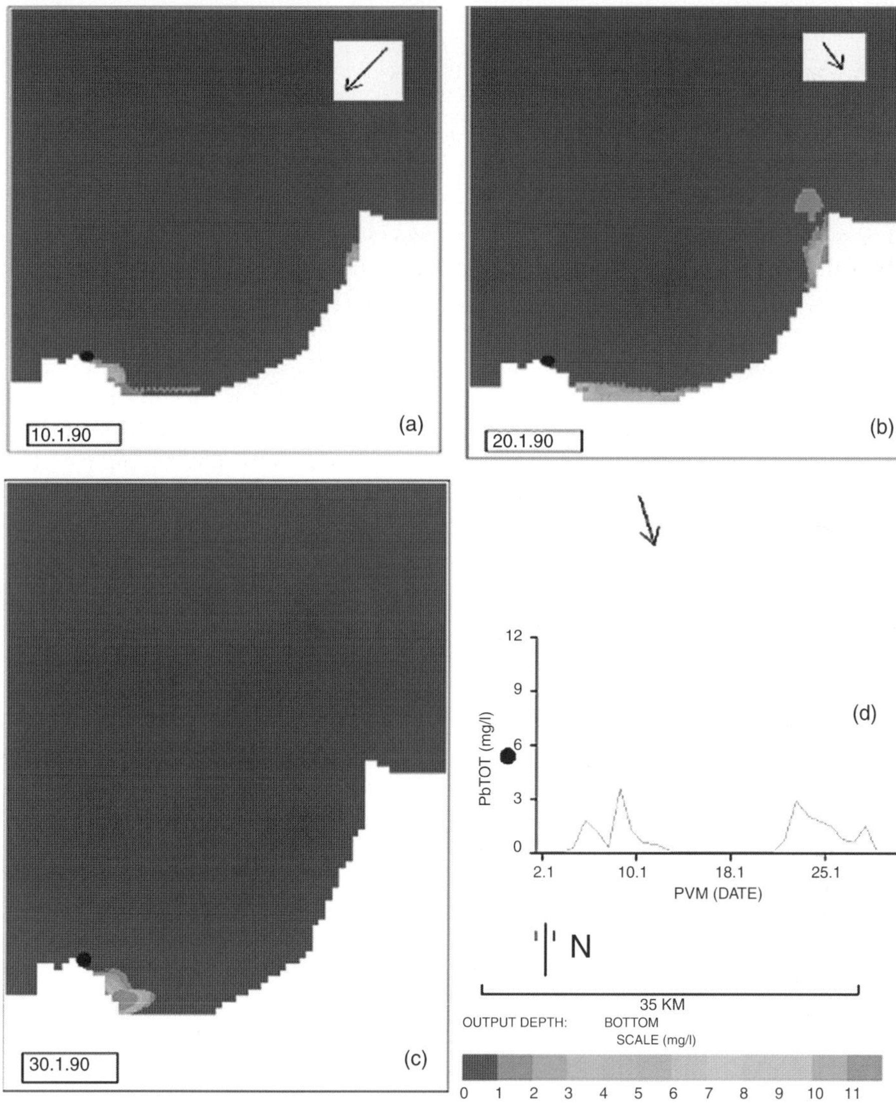

Figure 4. Distribution of PbTOT in Abu-Qir Bay during the winter.

erosion, water quality control, etc. should be continued to serve as database for coastal engineering, gas exploration and exploitation and harbors construction projects planned for the Bay.

4. The construction of industrial complexes along the coastline should not be built at erosional or depositional areas.
5. Development of a policy for the protection of submerged archaeological sites in a step to establish natural underwater museum.
6. Increase public awareness for the sustainable development of Abu-Qir Bay.

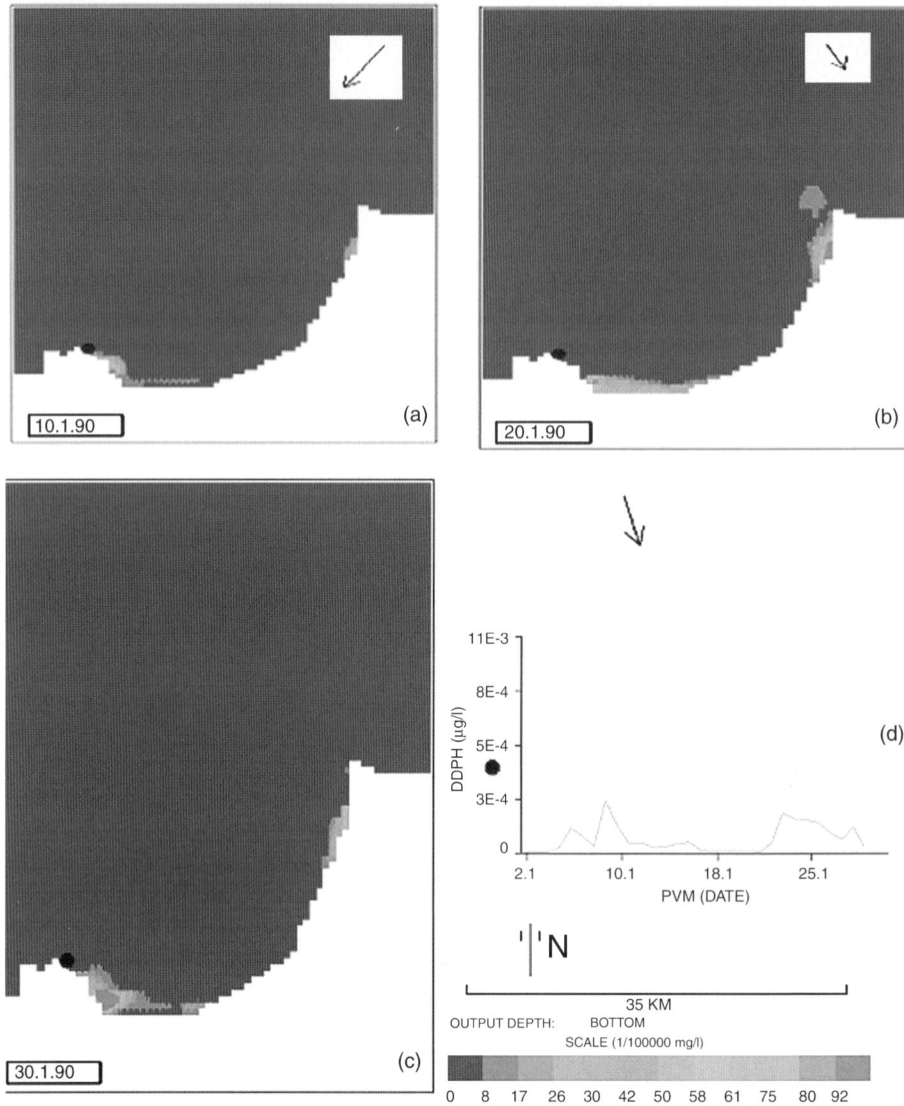

Figure 5. Distribution of DDPH in Abu-Qir Bay during winter.

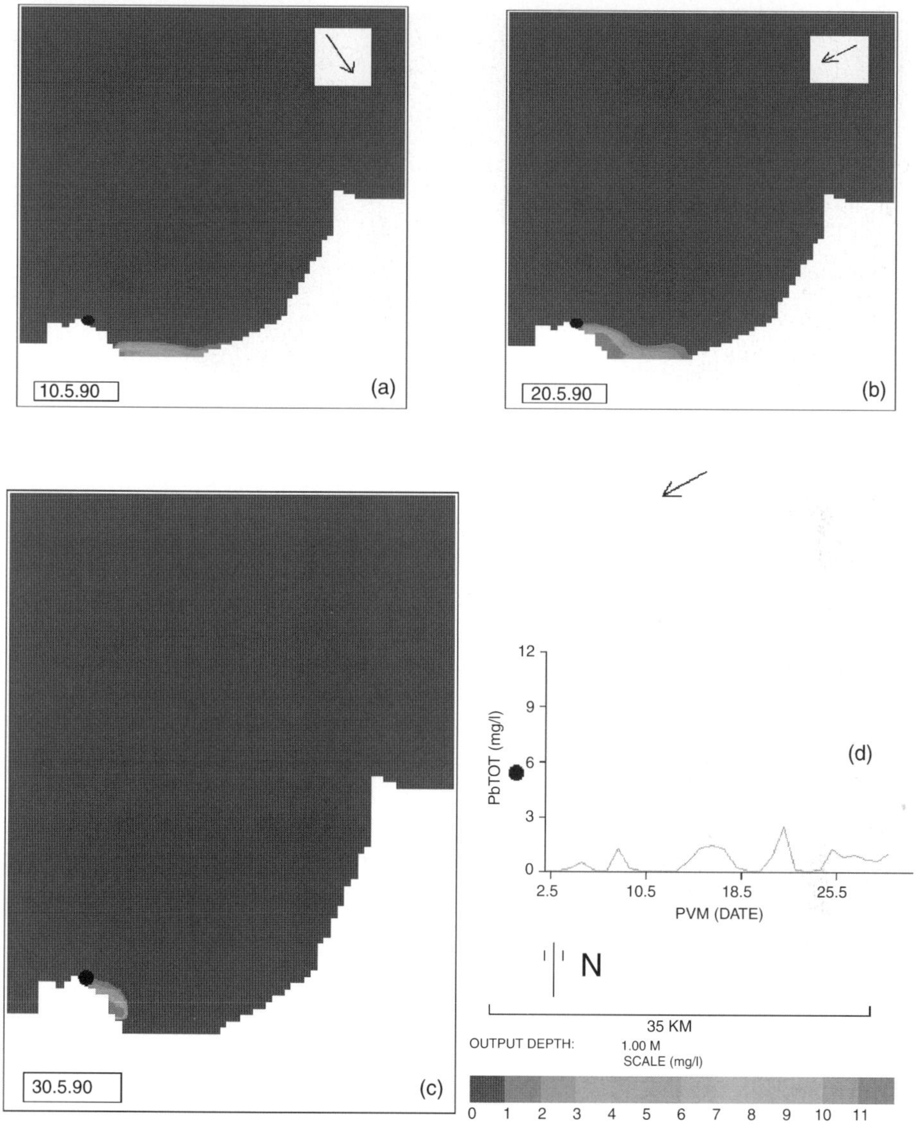

Figure 6. Distribution of PbTOT in Abu-Qir Bay during spring.

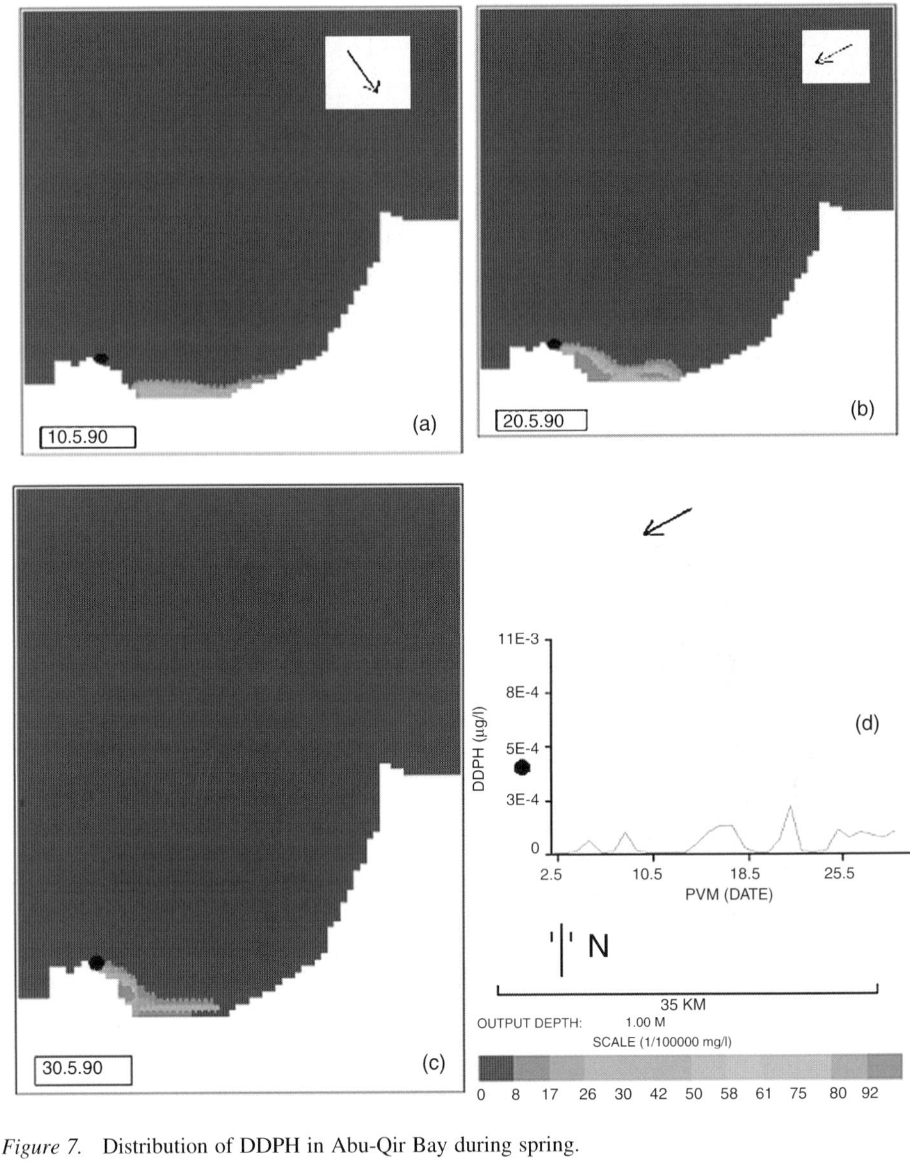

Figure 7. Distribution of DDPH in Abu-Qir Bay during spring.

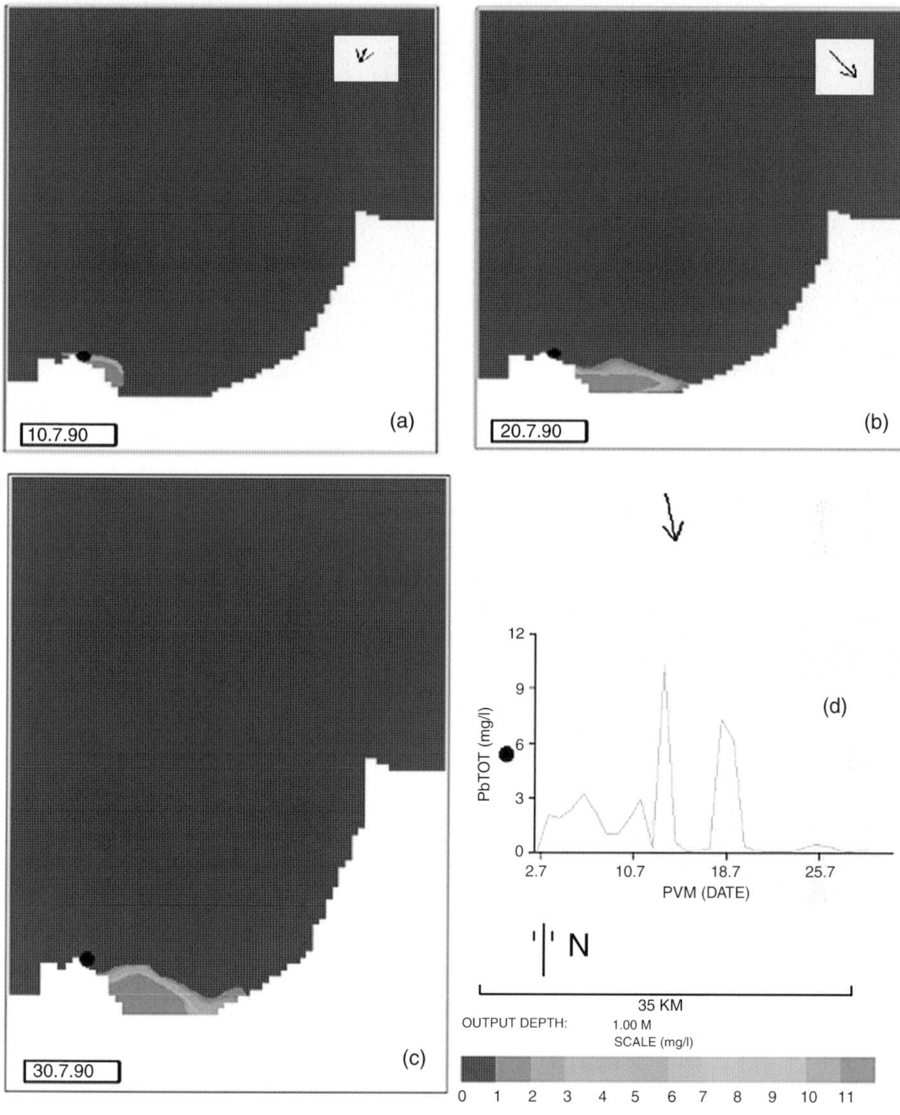

Figure 8. Distribution of PbTOT in Abu-Qir Bay during summer.

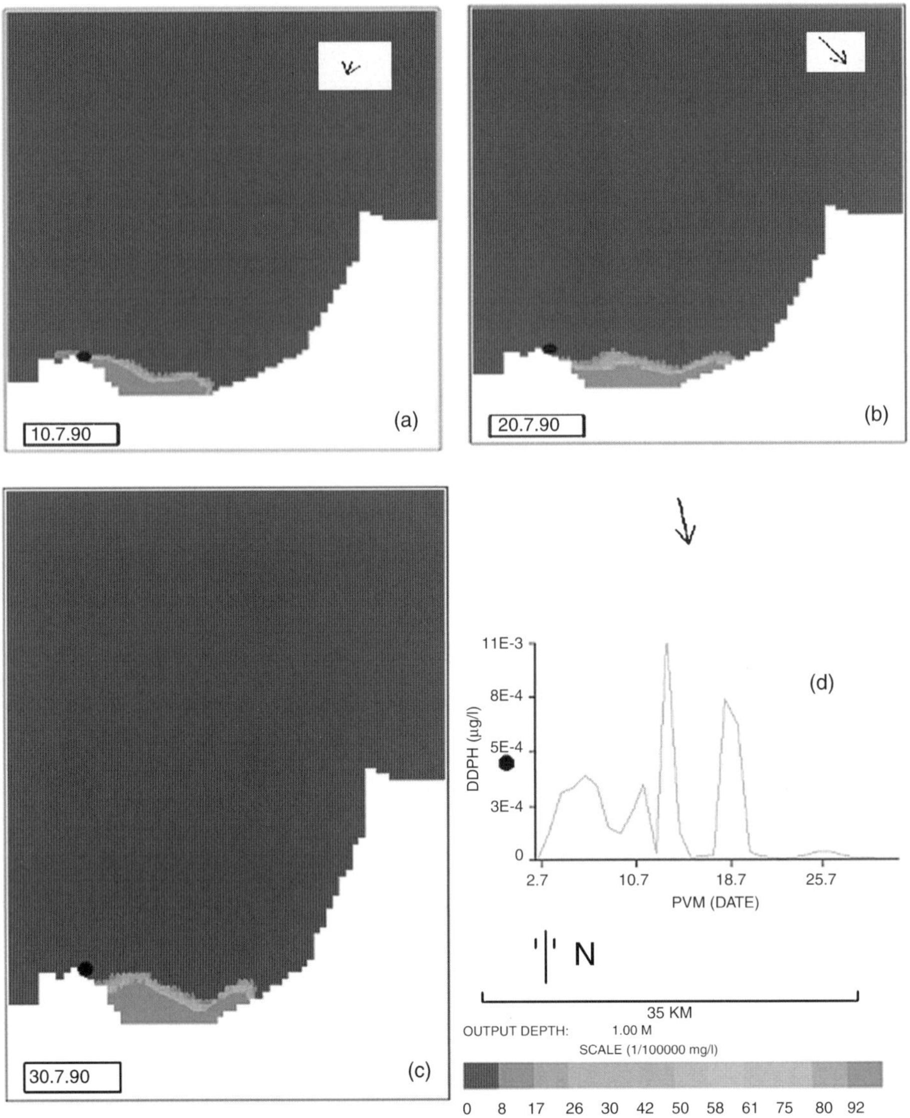

Figure 9. Distribution of DDPH in Abu-Qir Bay during summer.

References

Abdel-Moati, M.A., 2001. Long term impact of land base activities on the coastal waters of Alexandria beaches. MEDCOAST01, Hammamet, Tunisia, pp. 463–474, Oct. 22–24.

El-Haridi, N.M.G., 1994. Assessment of the potential release of selected heavy metals in the sediments of the Mahmoundia Canal. Ph.D. Thesis, Alexandria University, p. 239.

Halfon, E., Simons, T.J., Scherzer, W.M., 1990. Modelling the spatial distribution of seven halocarbons on Lake St. Clair in June 1984 using the TOXFATE model. J. Great lakes Res. 16, 90.

Kibar, N., 1997. Modelling transport in Beirut Bay, Lebanon. MEDCOAST97, Qawra, Malta, pp. 1145–1153, Nov. 11–14.
Koponen, J., 1991. Application of complex environmental numerical models. Doctor of Technology Thesis, Helsinki University of Technology.
Koponen, J., Alasaarela, E., Lehtinen, K., Sarkkula, J., Simbierowicz, P., Veps, H., Virtanen, M., 1992. Modelling the Dynamics of a Large Sea Area. Publications of Water and Environment Research Institute, No. 7, National Board of Waters and the Environment, Helsinki, Finland.
Meligy, M., 2000. Hydrographic Study on the area of Abu-Qir port in Abu-Qir Bay, M.Sc. Arab Academy for Science and Technology and maritime Transport.
Sarkkula, J., 1991. Application measuring and modeling water currents and quality as part of decision-making process for water pollution control. Ph.D. Thesis, Tartu University, Estonia.
Sarkkula, J., Virtanen, M., 1983. Application of Mathematical Flow Field Models in Finland. Swedish IHP Report no. 56, pp. 93–111.
Sarkkula, J., Koponen, J., Kuusisto, M., 1998. Modelled phytoplankton dynamics in the Gulf of Finland. Environ. Model. Software 13 (1998), 461–470.
Shabara, S.A.A., 2000. Monitoring of oil pollution in the marine environment along Alexandria coast. M.Sc. Thesis, Alexandria University, p. 147.
Simons, T.J., 1980. Circulation models of lakes and inland seas. Can. Bull. Fish. Aquat. Sci. 203, 145.
Soussi, N., Ennet, P., Koponen, J., Sarkkula, J., Ben Mustapha, S., Ben Mammou, A., Kokkila, T., 1995. Impact of phosphogyps waste in the Gulf of Gabes. MEDCOAST95, Tarragona, Spain, pp. 1334–1346, Oct. 24–27.
Virtanen, M., Koponen, J., 1985. Simulation of transport under irregular flow conditions. Aqua. Fenn. 15, 65–75.
Virtanen, M., Koponen, J., Helisten, S., Nenonen, O., Kinnunen, K., 1994. Principals for calculation of transport and water quality in strongly regulated reservoirs. Ecol. Model. 74, 103–123.
Younes, W.A.N., 1997. Nutrients and heavy metals in coastal waters of the Mediterranean Sea in front of Alexandria. M.Sc. Thesis, Alexandria University, Faculty of Science, p. 696.
Younes, W.A.N., Saad, M.A.H., Abdel-Moati, M.A.R., 1997. An ICZM approach: land-based sources of pollution and their effect on the submerged archeological sites of Alexandria. Proceedings of Third International Conference on the Mediterranean Coastal Environment, MEDCOAST97, Qawra, Malta, pp. 1375–1385, Nov. 11–14.
Younes, W.A.N., Romano, J.C., Arlhac, D., Bensoussan, N., 1999. Influence of rivers on adjacent coastal waters: observations on the Rhone and Nile rivers. MEDCOAST99, Antalya, Turkey, pp. 1674–1687, Nov. 9–13.

Oil Pollution and its Environmental Impact in the Arabian Gulf Region
M. Al-Azab, W. El-Shorbagy and S. Al-Ghais, editors

Chapter 9

Fate modeling of benzo(a)pyrene near a refinery outfall in the coastal water of United Arab Emirates

Mir Azzam[1], Walid El-Shorbagy[2,*] and Mohamed A. Karim[3]

[1]*Scientific Research Sector, United Arab Emirates University, P.O. Box 17555, Al-Ain, United Arab Emirates*

[2]*Civil & Environmental Engineering Department, United Arab Emirates University, P.O. Box 17555, Al-Ain, United Arab Emirates*

[3]*Chemical Engineering Department, United Arab Emirates University, P.O. Box 17555, Al-Ain, United Arab Emirates*

Abstract

A three-dimensional numerical modeling study is done to simulate the fate of benzo(a)pyrene, a carcinogenic poly aromatic hydrocarbon, in the dynamic coastal environment after its release from a refinery outfall. In the model, pollutants are distributed in the water, organic carbons, and sediments according to specific chemical laws. In the water column, it is subject to transportation, settling and losses. Processes of burial, resuspension, bioturbation, and losses regulate its kinetics in the sediment layer. The physical transport phenomena is governed by coastal currents induced by tides, wind, and water density gradients. These dynamics are extracted from a larger regional hydrodynamics model that resolves the flow system of the entire Arabian Gulf taking into account the oceanographic and meteorological forcing. The discharging concentration of the pollutant is 10 ppb in the refinery effluent. Results show concentration of more than 0.4 ppb of benzo(a)pyrene reported at radial distance of about 3 km from outfall. In the deeper water, the concentration is found less than 0.2 ppb.

1. Introduction

Wastewater is released in the coastal water from a refinery, which produces about 420,000 bpd of oil, in Ruwais on the eastern coastline of UAE. The coast, which accommodates a port for oil-gas tankers, is partially sheltered by an island and tidal flats (Fig. 1). The release of wastewater containing hydrocarbons has become a subject of concern in many parts of the UAE coast. Several aromatic components of the hydrocarbons are reported to be carcinogenic and persistent in nature as they enter the marine food chain or settle on the seabeds being adsorbed to the particulate materials. However, transformation and transport process of these hydrocarbons are substantially

*Corresponding author. *E-mail address:* walid.shorbagy@uaeu.ac.ae (W. El-Shorbagy).

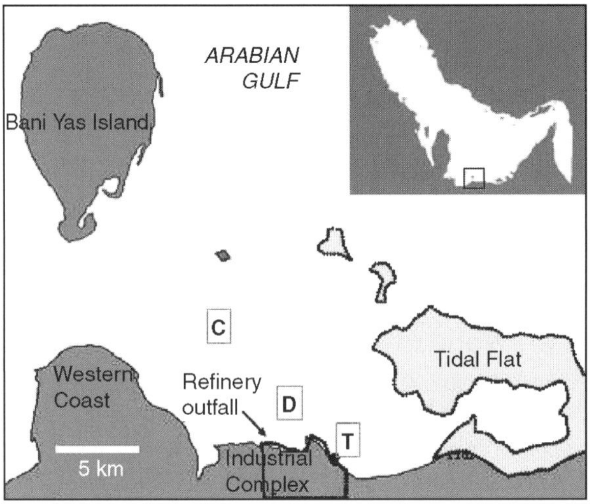

Figure 1. Study area showing the measurement points.

complex because of simultaneous influences of the physical, chemical and biological kinetics of the marine environment. Mathematical models can effectively track the fate of the chemicals pollutants in the nature on the basis of well-established chemical kinetics. Benzo(a)pyrene is one of such aromatic hydrocarbons that is usually found in the refinery wastewater.

In the present study, the transport and transformation of benzo(a)pyrene is modeled as the pollutant is released with refinery wastewater in the coastal water of Ruwais. The fate model is coupled with a calibrated hydrodynamic model.

1.1. Model description

Hydrodynamic modeling is needed to characterize the flow field conditions in the surface water. Such flow is forced by tide at the open boundaries, wind stress at the free surface, pressure gradients due to free surface gradients (barotropic) and density gradients (baroclinic) (Azzam et al., 2004a).

The model employs the classical mass and momentum principles to solve for the shallow water equations in three dimensions. The equations are formulated in rectilinear co-ordinates transformed into orthogonal curvilinear co-ordinates on horizontal plane and sigma grid system used to resolve the vertical axis. Source and sink terms are included in the equations to model the discharge and withdrawal of water.

The dynamics of the micropollutants are resolved by implementing advection–dispersion equation. The equation is in fact a mass balance, which is formulated in terms of finite differences. The mass balance equations provide the change of a concentration over a known time step in a compartment. They are different for the water column and sediment layer as follows:
For water:

$$(\Delta C_t/\Delta t) = \text{loads} + \text{transport} + \text{resuspension} - \text{settling} - \text{losses} \qquad (1)$$

For sediment:

$$(\Delta C_t/\Delta t) = \text{transport} - \text{settling} - \text{resuspension} - \text{burial} + \text{digging} + \text{bioturbation} - \text{losses} \tag{2}$$

where C_t is the total concentration of micropollutant.

It has to be noticed that the micropollutant concentration in this case can be described by the total concentration (sum of dissolved and particulate concentration), the total particulate concentration, or by the total dissolved concentration for each water and sediment compartment. The particulate and dissolved concentrations are derived from the total concentration and respective fractions. The later are calculated from partitioning formulations.

The adsorbed fractions in the water column are subject to settling and the fractions in the sediments are subject to resuspension. The adsorbed fractions in the sediments can also be removed from the modeled part of the water system by burial.

Adsorption to the particulate matter is described by means of equilibrium isotherms or kinetic partitioning, on the basis of partitioning coefficient. The fractions can be shown in the following form:

$$C_t = (f_{im} + f_{poc} + f_{alg} + f_d)C_t \tag{3}$$

The bracketed terms are the fraction of micropollutant to inorganic matter, dead particulate matter, adsorbed to algae and dissolved amount, respectively.

The fractions are determined by using different partitioning coefficients (K_{dim}, K_{dpoc}, K_{dkalg}, and K_d) estimated either by an equilibrium approach or by sorption kinetics. The equilibrium method uses the following relation to estimate the partition coefficients:

$$K_d = f(C_{p'}, C_s, C_d) \tag{4}$$

where $C_{p'}$ is the particulate concentration of a micropollutant (kg/m^3), C_s, the concentration of particulate matter (kg/m^3), C_d, the dissolved concentration of micropollutant (kg/m^3). The sorption kinetic method uses the following relation to estimate the rate of adsorption and desorption:

$$R_{sorp} = K_{sorp}(C_{pe'} - C_{p'}) \tag{5}$$

where $C_{pe'}$ is the equilibrium particulate concentration (kg/m^3), K_{sorp}, the first-order kinetic constant for adsorption and desorption (kg/m^3/s). There are several extra processes important in micropollutant dynamics in water. These include: volatilization, biodegradation, photolysis, hydrolysis, and overall degradation. These processes are formulated as first-order kinetics. However, some of them are pseudo first-order since they are functions of other state variables incorporated in the model as forcing functions.

Volatilization is calculated according to the two-film theory. The volatilization rate is equal to:

$$R_{vol} = -F_{vol}/H = (K_{vol}C_d)/H \tag{6}$$

where C_d is the dissolved concentration (kg/m^3), F_{vol}, the volatilization mass flux (kg/m^2/s), H, the thickness of the upper water layer (m), K_{vol}, the overall transfer coefficient for volatilization (m/s), R_{vol}, the volatilization rate (kg/m^3/s).

All loss processes are basically described by the equation: $R = -K \times C_t$ (7)

where C_t is the total concentration of micropollutant (kg/m^3), K, the pseudo first-order kinetic constant for biodegradation, photolysis, hydrolysis, and overall degradation (s^{-1}), R, the rate of biodegradation, photolysis, hydrolysis, and overall degradation (kg/m^3/s).

1.2. Model setup

The model employs curvilinear grid for the horizontal plane. The numerical grid consists of 31,331 cells with side dimensions 231 and 163 along east–west and north–south directions, respectively. The vertical dimension is modeled in sigma co-ordinate with five layers. The smallest grid size is about 40 m × 40 m. The central part of the area is deep with a depth up to 18 m and the areas at the east and the west are very shallow. The model is nested inside a larger regional model (Elshorbagy et al., 2004a) developed for the entire Arabian Gulf to adopt boundary flow data. The regional model has its boundary at the Straits of Hormuz. Several model parameters have been employed from a recent study (Elsorbagy et al., 2004b).

Initially, the hydrodynamic model is calibrated by adjusting model parameters within practical ranges to attain agreement with measured hydrodynamic data. Comparison of the measured and simulated water level data at location T is shown in Figure 2 indicating a satisfactory agreement. The depth-averaged eastward and northward currents computed by the model are compared with measured data at location C (Fig. 3). The match of the estimated eastward and northward current with observation appears to be satisfactory.

The chemical fate model is setup to simulate the dynamic distribution of benzo(a)pyrene as it is released from the refinery outfall. The pollutant concentration of the refinery effluent, which is flowing out at the rate of 1 m^3/s, is 1 ppm. The physical flow condition estimated by the hydrodynamic model is dynamically incorporated by the fate model. The initial and boundary concentration of benzo(a)pyrene is considered to be zero. The environmental fate parameters are obtained from Mackay et al. (2000). Organic carbon and phytoplankton concentration values are obtained from field measurement. Values and parameters used in the model are presented in Table 1.

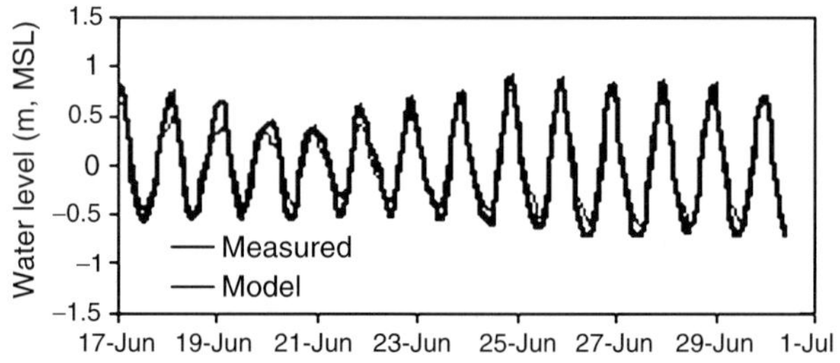

Figure 2. Comparison of measured and simulated tide at location T.

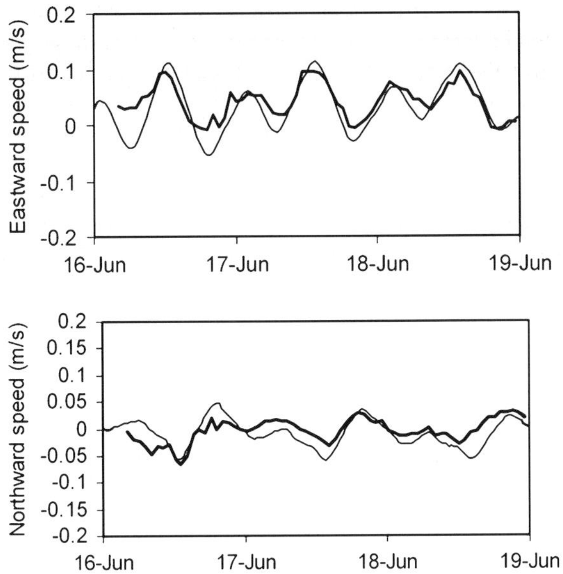

Figure 3. Comparison of measured and simulated current components in eastward and northward directions.

2. Results

2.1. Hydrodynamics

The model output shows that a net anticlockwise circulation is generated in the northwestern basin area following the shore of the Bani Yas Island. Almost no mean flow is observed in the central and coastal areas. A mean current enters the area from west through the opening between the Bani Yas Island and the main land and the balancing outflow occurs through the north. A number of eddies are formed close to the north and

Table 1. Model parameters and necessary input values for the fate model.

Constants/Parameters	Values
Log partitioning coefficient to particulates (l/kgC)	7.03
Log partitioning coefficient to phytoplankton (l/kgC)	7.03
Henry's law constant (Pa/m^3/mol)	0.0089
Thermodynamic constant	12
Concentration in atmosphere (g/m^3)	0.00001
Environmental half-life (overall degradation) d	3.2
Temperature coefficient for degradation	1.07
Efficiency DOC relative to POC	0.18
Mean POC (g/m^3)	0.27
Mean carbon in algae (g/m^3)	0.0046
Mean DOC concentration (g/m^3)	2.45

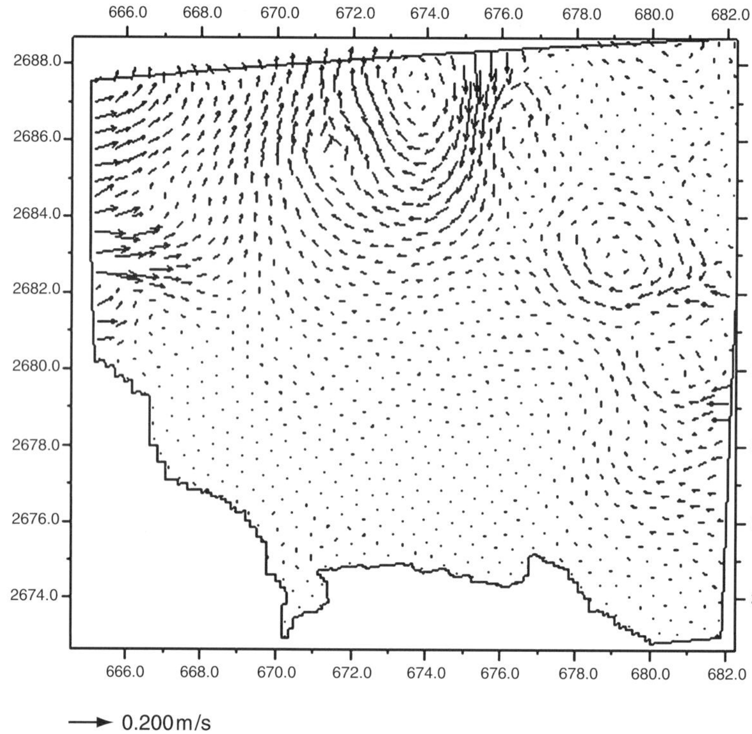

Figure 4. Simulated mean flow pattern in the study area.

east boundaries. Details of the model results are presented somewhere else (Azzam et al., 2004b) (Fig. 4).

2.2. Pollutant dynamics

Concentration of benzo(a)pyrene both in dissolved and adsorbed phase near the desalination intake is shown in Figure 5. Chemical concentration increases almost steadily for one week and then oscillates with tidal spring neap cycle with a period of

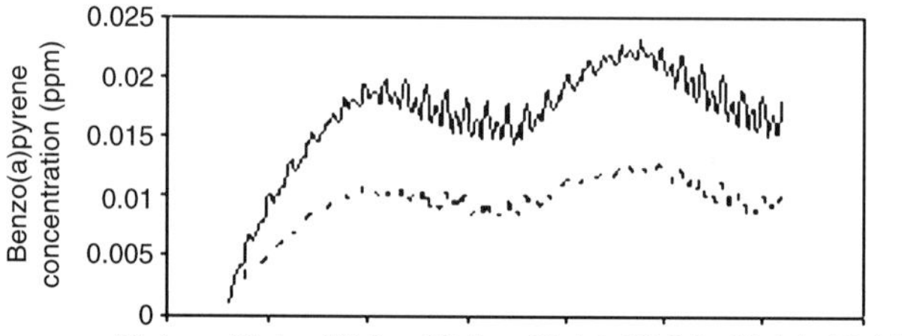

Figure 5. Time-dependent dissolved and adsorbed benzo(a)pyrene concentration near desalination plant intake point (D).

almost 14 days with respect to an apparent mean level of 0.016 $\mu g/m^3$ for dissolved phase and 0.01 $\mu g/m^3$ for adsorbed phase. The concentration is lowered during the spring tide as the tidal flux rises increasing the mixing in the coastal areas (Azzam et al., 2004b). With the reduction of tidal flux as well as the mixing rate during the neap tide, the concentration of the pollutant tends to increase until the advent of next spring tide. Similar oscillating pattern is also found for the adsorbed pollutants. Under the influence of diurnal tide, dissolved concentration fluctuates with a range of up to 0.004 $\mu g/m^3$ during spring tide, while the range reduces significantly during the neap tide. Daily fluctuation of adsorbed concentration both in the spring and neap tides are insignificant.

Model simulated distribution of dissolved benzo(a)pyrene in the study area after 30 days of continuous discharge from the refinery is shown in Figure 6. The concentration contours are almost radially distributed around the discharging outfall with sign of relatively higher rate of westward transport. Results show concentration of more than 30 ppb of benzo(a)pyrene outside the radial distance of about 2 km from outfall. In the deeper water concentration is found less than 20 ppb. Distribution of chemical adsorbed to particulates is shown in Figure 7 representing the scenario after 30 days of refinery waste discharge. In general, the adsorbed concentration is found almost half of the dissolved

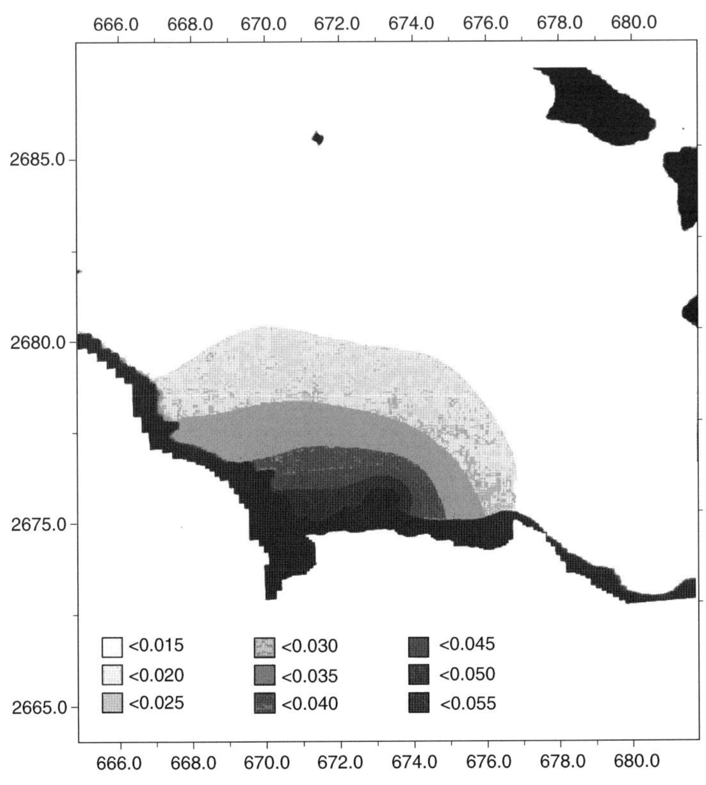

Figure 6. Distribution of dissolved benzo(a)pyrene after 30 days of continuous wastewater discharge from the refinery.

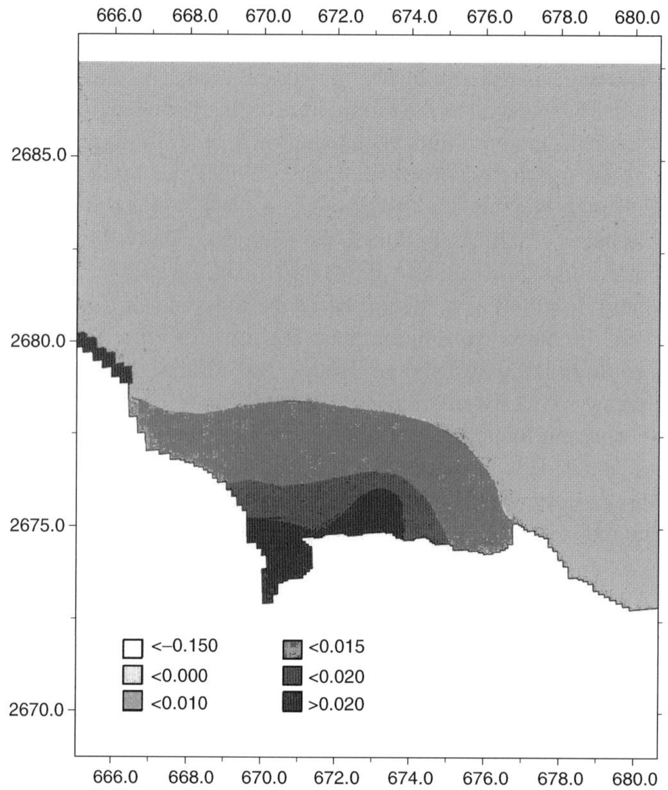

Figure 7. Distribution of adsorbed benzo(a)pyrene after 30 days of continuous wastewater discharge from the refinery.

concentration but the overall distribution trend shows similar pattern. Outside the radial distance of about 2 km from outfall, the adsorbed concentration is found to be 10 ppb.

References

Azzam, M.H., Elshorbagy, W., Ichikawa, T., Terasawa, T., 2004a. A three dimensional model application to study the residual flow in the Arabian Gulf. Submitted to the Journal of Waterway, Port, Ocean and Coastal Engineering.

Azzam, M.H., Elshorbagy, W., Nakata, K., 2004b. Three dimensional modeling of the Ruwais coastal area of United Arab Emirates. Paper Submitted to Journal of Port, Coast, Waterway and Ocean Engineering.

Elshorbagy, W., Azzam, M.H., Taguchi, K., 2004a. Hydrodynamic characterization and modeling of the Arabian Gulf. Submitted to the Journal of Waterway, Port, Ocean and Coastal Engineering.

Elshorbagy, W., Azzam, M.H., Ichikawa, T., 2004b. Temperature – salinity modelling for Ruwais Coast, UAE. Submitted to ASCE Journal of Port, Way, Coastal and Ocean Engineering.

Mackay, D., Shiu, W.Y., Ma, K.C., 2000. Physical–Chemical Properties and Environmental Fate Handbook. Chapman and Hall, London.

Oil Pollution and its Environmental Impact in the Arabian Gulf Region
M. Al-Azab, W. El-Shorbagy and S. Al-Ghais, editors

Chapter 10

The role of oil-sediment aggregation in dispersion and biodegradation of spilled oil

Ali Khelifa[1,*], Paul S. Hill[1] and Kenneth Lee[2]

[1]*Department of Oceanography, Dalhousie University, Halifax, NS, Canada, B3H 4J1*
[2]*Center for Offshore Oil and Gas Environmental Research, Bedford Institute of Oceanography, Fisheries and Oceans Canada, P.O. Box 1006, 1 Challenger Drive, Dartmouth, NS, Canada B2Y 4A2*

Abstract

It is now widely acknowledged that suspended sediments have strong influence on the fate and transport of spilled oil into the aquatic environment. This chapter presents a brief summary of the state of knowledge on how suspended sediments contribute, via the process of heteroaggregation oil-sediment, to dispersion of spilled oil and its biodegradation. Effects of environmental parameters such as salinity, sediment concentration and size as well as oil type on this process are discussed. The study shows that this recently recognized mechanism by the oil spill remediation experts has been, in fact, discovered earlier by the production community in the beginning of the last century. Discussion on real application is presented to show how this process can enhance considerably natural recovery of oiled shorelines when the method of "surf washing" is applied. An outline of research required to improve our understanding of the process is finally presented.

1. Introduction

Oil spilled in aquatic environments is subjected to complex transformations immediately after it enters the water. The progression, duration, and results of these transformations depend on the properties and composition of the oil, spill conditions, as well as environmental parameters. Dispersion and splitting of oil slicks into small droplets is, for instance, one of the key transformations that affect the fate and transport of spilled oil. Small oil droplets are easily dispersed by hydrodynamic flows and are less harmful to aquatic environments as they are more easily biodegraded than oil slicks (Bragg et al., 1994; Weise et al., 1999). Breaking waves are acknowledged as the key factor that controls the transformation of oil slicks into small oil droplets and dispersion into the water column. Nevertheless, resulting oil droplets are instable. They can recoalesce after

[*]Corresponding author at present address: Environmental Technology Centre, Environment Canada, 335 River Road, Ottawa, ON, Canada K1A OH3. *E-mail address:* ali.khelifa@ec.gc.ca (A. Khelifa).

collisions and float to the surface (due to the buoyancy of the oil) to form new slicks when the wave energy weaken.

Very simple laboratory tests performed by researchers in the 1940s and 1970s revealed, however, that when an oil slick is introduced at the water surface in small containers filled with clay–water mixture and shaken at a constant energy, dispersed oil droplets do not reform oil slicks (Poirier and Thiel, 1941; Bassin and Ichiye, 1977; Huang and Elliot, 1977; Zurcher and Thuer, 1978). Various interpretations have been proposed to explain this intriguing observation. Most of them were, however, based on visual observation of the process. The significance of this process to dispersion of spilled oil remained ignored for decades until the huge Exxon Valdez oil spill occurred in 1989 in Alaska. Continuous monitoring of the contaminated sites years after the spill showed that many sections of oiled shorelines, such as Northwest Bay and Herring Bay, cleanup naturally (Bragg and Yang, 1990, quoted in Owens, 1999). Further investigations of the contaminated sites performed in the early 1990s showed abundance in the water column of microsized oil droplets coated with fine clay particles. The hypothesis of clay–oil flocculation has then been proposed as one of the key mechanisms behind the natural cleansing of the oiled shorelines (Bragg and Owens, 1994; Owens et al., 1994; Bragg and Yang, 1995). Since then, various terminologies have been used to refer to this process, which represents in fact a heteroaggregation between clay and oil particles. Among these terminologies are *mineral-stabilized droplet*, *Clay–oil flocculation* (Bragg and Owens, 1994; Bragg and Yang, 1995), *Oil and fine-particle interaction* (Owens 1999), and more frequently *oil-mineral aggregate* (OMA) (Lee et al., 1998). Even if the actual knowledge of the process suggests that the later terminology (OMA) is not necessarily the appropriate one, as aggregation between oil droplets and non-mineral particles were widely observed, we shall use it hereafter.

The goal of this study is to provide a summary on the state of knowledge of OMA formation and its role in dispersion and biodegradation of spilled oils. The review is followed by an outline of related topics that are still under investigation and need further considerations in the future to improve our quantitative understanding of this natural process.

2. Mechanism of OMA formation

Formation of OMA occurs naturally when oil and suspended solid particles are present in a turbulent aqueous medium. Basically, in this process fine sediment grains adsorb on the surface of the droplet as grains or flocs to form a physical barrier around the droplets (Fig. 1). This barrier plays a triple role: it enhances the stability of droplets in the water column by preventing re-coalescence of droplets and adhesion of droplets to shorelines; it enhances the density of droplets to make them neutral or slightly negatively buoyant, which facilitates their dispersion into a larger water mass; and more importantly it enhances the biodegradation rate of the spilled oil as the water–oil interface increases considerably when an oil slick is split to microsized droplets (Lee et al., 1997; Jézéquel et al., 1999; Weise et al., 1999). This vision of the OMA process is generally shared by researchers dealing with oil spill remediation problems and, as mentioned above, is relatively recent (Bragg and Owens, 1994; Bragg and Yang, 1995). More recently, the

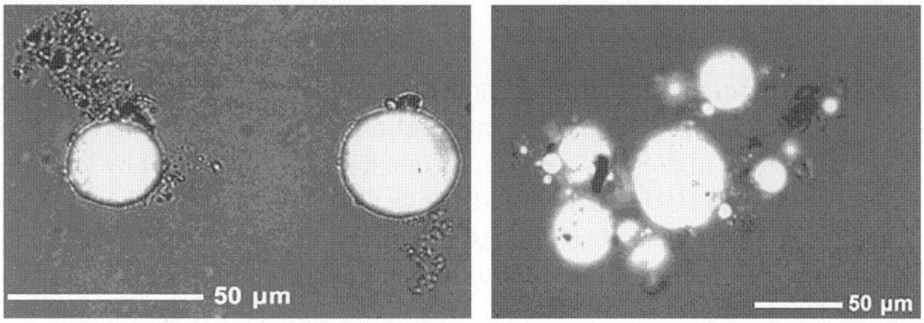

Figure 1. Examples of droplet OMA (left) and multidroplet OMA (right) observed in laboratory using epi- fluorescence microscopy (Stoffyn-Egli and Lee, 2002): the oil appears bright (fluorescent) and the mineral particles are darker surrounding the droplets.

same group suggested that OMA formation may play a major role in the natural cleaning of oiled shorelines and may be the basis for the development of oil spill countermeasure technologies (Lee et al., 1997, 1998, 2003a,b; Lee and Stoffyn-Egli, 2001; Wood et al., 1998; Jézéquel et al., 1999; Owens, 1999; Sergy et al., 1999; Guyomarch et al., 1999, 2002; Hill et al., 2002; Kepkay et al., 2002; Khelifa et al., 2002, 2003a,b; Le Floch et al., 2002; Muschenheim and Lee, 2002; Omotoso et al., 2002; Guénette et al., 2003; Owens and Lee, 2003).

From a different perspective, the OMA process has been reported as an oil production problem by the oil production community, more precisely during the production phase called "dedusting." It is general knowledge within this group that solid particles are capable of stabilizing the droplets of one fluid in another.

Indeed, Ramsden (1903) first observed emulsions stabilized by the presence of solid or highly viscous matter at the interface of two liquids. Pickering (1907) discovered that finely divided solid particles wetted more by water than oil could act as an emulsifying agent for oil-in-water emulsions. Since then, numerous investigations have been focused on the stability of solids-stabilized emulsions (e.g. Menon and Wasan, 1988; Yan and Masliyah, 1995; Binks and Lumsdon, 2000a,b).

To date, experimental and theoretical contributions from the two groups show that formation of OMA is a process that occurs naturally after oil spills or during oil production. OMA form readily if the conditions are favorable. The following section discusses main factors controlling OMA formation.

2.1. *Controlling factors*

Fundamental requirements for OMA to form are presence of dispersed oil droplets, fine sediment grains, and sufficient mixing energy to keep oil and sediment particles in suspension and to insure collision between them. Among other factors that are recognized affecting concentration of OMA are physical properties of sediment such as size, density, composition and concentration, oil properties such as viscosity, composition, density and concentration, and environmental conditions such as turbulence, temperature and salinity. However, the significance of most of these factors on OMA formation is still not understood quantitatively.

2.1.1. Effects of salinity

Early investigators from oil remediation community reported that OMA formation is due more to the electrolytic flocculation of the clay particles than to any affinity between the oil and sediment particles. Oil droplets are trapped into the network formed by large sediment flocs (Meyers and Quinn, 1973; Bassin and Ichiye, 1977). Salinity of the aqueous phase and the mineralogy of suspended sediments were then considered as the key parameters controlling the process. Further in the early and mid 1990s, researchers concluded also, from laboratory and field tests, that formation of OMA enhances when mineral fines with size less than about 5 μm are abundant in the aqueous phase (Payne et al., 1989; Bragg and Owens, 1994; Bragg and Yang, 1995). But, recent laboratory studies on North Slope Rivers in Alaska (McCourt and Shier, 1998, 2000; Hoffman and Shier, 1999) and the field study conducted by Lee et al. (2002) following the OSSA II oil spill in Bolivia River showed that formation of OMA also occurs in freshwater. A possible explanation to these observations was attributed to the presence of cations in freshwater other than Na^+ and to the type of the reacting sediments. Our recent laboratory study (Khelifa et al., 2003a, in press) showed strong effects of salinity on OMA formation. We speculated in that study that observed effects of salinity on OMA formation are a consequence of combined effects of salinity on the adsorption of clay particles on the surface of oil droplets, because of the diminution of electrokinetic charges of both oil droplets and clay particles, and by the flocculation of clay particles. The following relationship was then established to predict the effects of salinity on the concentration of OMA-stabilized oil:

$$N_* = \frac{(S_*^{1.97} + 0.01)}{(S_*^{1.97} + 0.12)} \qquad (1)$$

where $N_* = N_t/(N_t)_{max}$ represents the normalized number concentration of OMA-stabilized oil droplets and $S_* = S/S_{cas}$ is the normalized salinity in which S is the salinity and S_{cas} is the critical aggregation salinity at which the observed rapid increase of the number concentration N_t of stabilized droplets with salinity ends and where N_t reaches its maximum value $(N_t)_{max}$. In other words, beyond this salinity there is no substantial increase of N_t. Overall, the data and Eq. (1) showed that the critical condition under which the formation of OMA increases rapidly is expected when S_* varies between 0.1 and 1 (Fig. 2). However, a little is known about the variable S_{cas}. Our previous study showed that this parameter is affected by sediment type and, to a lesser extent, by oil type. It varies between 1.2 and 3.5 ppt, but might be as large as 35 ppt and as small as 0.2 ppt.

2.1.2. Effects of sediment size

Size of sediment grains is also recognized to play a major role in OMA formation. Based on their investigation of five different oiled sites, Bragg and Owens (1994) suggested that sediments with particle size approaches the clay range (≤2 μm) are favorable candidates for OMA formation. They reported, however, that because particle size usually follows a normal distribution, even fines that have a mean particle diameter of 6–10 μm likely will contain sufficient clay-size particles to support efficient OMA formation. According

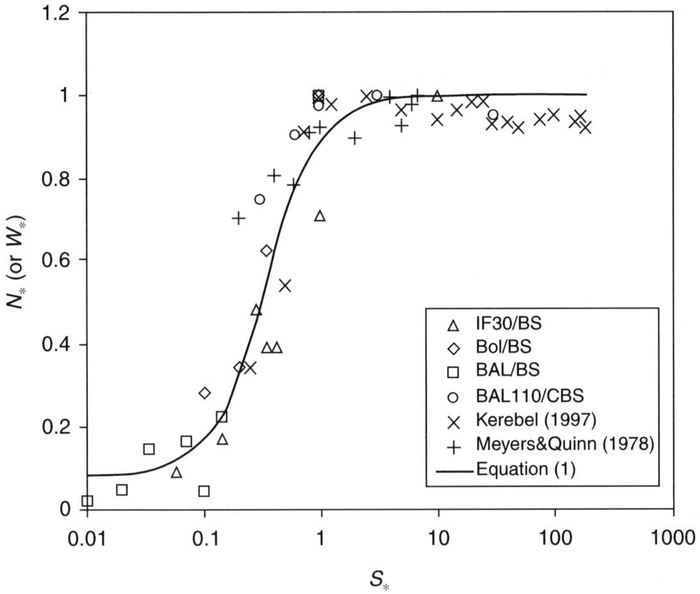

Figure 2. Variations of the concentration ratio N_* (or W_*) with salinity ratio S_*. W_* represents the mass ratio.

to Lee et al. (1997) field-sampled OMA following the Sea Empress oil spill (United Kingdom) contain a variety of mineral particles, ranging in size from ≈ 15 μm to less than 1 μm. Results from the recent laboratory study conducted by Ajijolaiya (2004) showed that the concentration of OMA-stabilized oil droplets decreases by a factor six when sediment-grain size increases from 0.5 to 16 μm. This is also supported by our recent modeling results (Khelifa et al., 2004a). Our results showed that the magnitude of the size effect is oil-dependent.

2.1.3. Effects of sediment concentration

Different laboratory studies have shown the significance of sediment concentration on OMA formation (McCourt and Shier, 1998, 2000; Guyomarch et al., 1999, 2002; Ajijolaiya, 2004). The magnitude of the effect is oil and salinity dependant. Overall, experimental data from these studies showed that OMA-stabilized oil concentration is very small at low sediment concentrations. As sediment concentration increases, OMA-stabilized oil concentration increases rapidly to reach a maximum. Further increase of sediment concentration produces insignificant effects on OMA formation. These experimental evidences are supported by our recent numerical simulations (Khelifa et al., 2003b, 2004b). In a companion paper (Khelifa et al., 2004b) we showed from numerical simulations that minimum sediment concentration (MSC) exists below which formation of OMA does not occur. This MSC is sediment size and oil type dependant and varies from about 0.5 to 200 mg/l. McCourt and Shier (2000) found that the impact of OMA formation on a spill in seven glacier-fed rivers in south-central Alaska would be small when the concentration of suspended solids is below 125–150 mg/l. In a related study, Hoffman and Shier (1999) reported that total suspended solids concentrations ranging from 1 to 190 mg/l were found insufficient to form OMA.

2.1.4. Effects of oil type

Existing evidences about the effects of oil type on OMA formation showed that oil viscosity and asphaltenes-resins content play a major role. Oil viscosity controls droplet formation, while asphaltenes-resins content appears to enhance adhesion between oil droplets and sediment particles (Khelifa et al., 2002). Bragg and Owens (1994) conducted an extensive study on OMA formation. Sediments and oiled sediment samples from five spills (Arrow 1970 in Nova Scotia in Canada, Metula 1974 in Straits of Magellan in Chile, Baffin Island Experiment 1980 in Cape Hatt NWT in Canada, Nosac Forest 1993 in Washington in USA, and Fred Bouchard 1993 in Florida in USA) were collected and analyzed. Both *in situ* and laboratory flocculated aggregates were observed under the microscope to confirm the presence of oil into the aggregates and to measure their size. They reported that oil containing no, or negligible polar hydrocarbons (NSO polars consisting of nitrogen, sulfur, and oxygen heteroatoms) do not flocculate with mineral fines. Two tests were performed to show such affirmation. In the first one the authors consider a pure alkane, hexadecane $[CH_3(CH_2)_{14}CH_3]$ and test it for flocculation against suspended fines from Smith Glacier and various Arrow spill locations. They observed no flocculation, and after extensive shaking in a test tube, the oil merely coalesced into large droplets with no mineral fines incorporated. In the second test, weathered Exxon Valdez oil was separated into saturate, aromatic, polar, and asphaltenes fractions by high performance liquid chromatography. The saturate and aromatic fractions were recombined to form non-polar oil, and this was tested for flocculation with Smith Glacier fines and various Arrow spill site fines. They observed again no flocculation. The authors reported, however, that almost any natural crude oil or residual refined oil likely will have sufficient polars to cause flocculation with fine particles (providing that its viscosity is sufficiently low).

We have shown in a previous laboratory study (Khelifa et al., 2002) that the effects of oil type as well as the effect of temperature can be predicted with the following relationship:

$$\frac{W_o}{W_{ar}} = 0.3\ e^{3.23(\mu_d/\mu_c)^{-0.22}} \tag{2}$$

where W_o is the ratio between the mass of OMA stabilized oil stabilized and the initial mass of oil introduced in the system, W_{ar} is the asphaltenes-resins content and μ_d and μ_c represent viscosities of the droplets and the continuous phase, respectively. The study showed that the effect of temperature on OMA formation is significant and is represented essentially by the resulting modification of oil viscosity. Low temperature reduces OMA formation. McCourt and Shier (1998) reported similar effects on OMA formation when the temperature decreases from 13–15°C to 2°C.

2.2. Time scale for OMA formation

2.2.1. Laboratory observations

Few laboratory and theoretical studies have addressed the kinetics of OMA formation. Laboratory experiments showed that formation of OMA occurs rapidly (Delvigne, 1987; Delvigne et al., 1987; Payne et al., 1989, 2003; McCourt and Shier, 1998, 2000; Guyomarch et al., 1999, 2002). In the majority of the experiments, OMA formed during the first 10 min when oil-sediment are mixed in small containers filled with water. The samples were, however, mixed in most of the cases at very high energy using hand shaker (Bragg and Owens, 1994), rotating shaker (Guyomarch et al., 1999), and oscillating shaker (McCourt and Shier, 1998, 2000). As far as the authors know, only two systematic investigations on the kinetics of OMA formation were performed in laboratory. The first one was conducted by Delvigne et al. (1987) using a grid column and the second study was performed by Payne et al. (1989) using a propeller-based reactor. Both studies showed that most of the OMA form during the first 10–20 min of the reaction. Data from Delvigne et al. (1987) are shown in Figure 3.

2.2.2. Theoretical modeling

Assuming that collision between sediment and droplet particles is controlled mainly by turbulent fluctuations, Hill et al. (2002) used the population balance equation to derive an

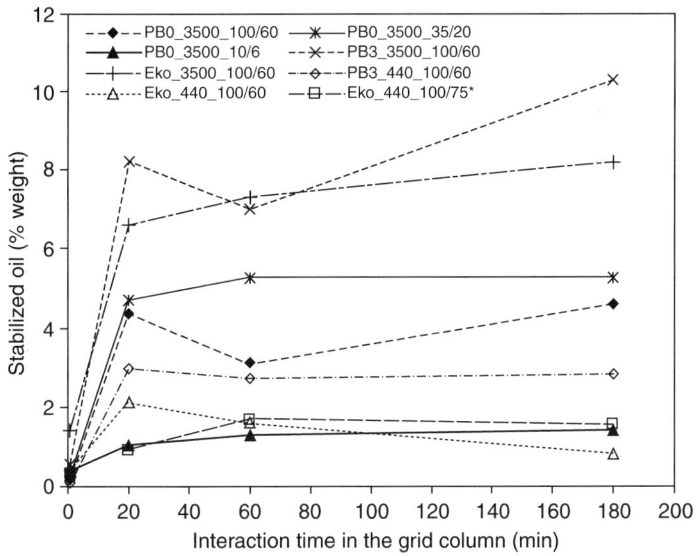

Figure 3. Variations of OMA-stabilized oil concentration (normalized with the initial concentration of oil) with the reacting time (data from Delvigne et al., 1987). The data were obtained with different oils and waddensea silt. The legend is such that the first letters represent the initial of the reacting oil: PB0, PB3 are for unweathered and 3 days weathered Prudhoe Bay crude (Alaska) and Eko for Ekofisk crude (North Sea). The following number represents the turbulent kinetic energy dissipation rate in (J/s m^3) and the last term represents the oil to sediment concentration ratio.

empirical relationship (Eq. 3) to calculate the time scale for OMA formation. Using this relationship, they performed several runs considering different conditions and concluded that the likely time scale for OMA formation varies between 5 min and 24 h. They reported that this range is consistent with the observed efficacy of the sediment relocation procedure (discussed in Section 3), as application of this procedure to oiled beaches results in cleaning them after few tide cycles (Lunel et al., 1995, 1996).

$$t_c = -\frac{\ln\left(1 - \frac{2\pi}{\sqrt{3}}\left(\frac{D_o}{D_s}\right)^2 \frac{N_o}{N_s(0)}\right)}{\beta} \tag{3}$$

with β given by

$$\beta = 0.16\alpha_{os}(D_s + D_o)^3(\varepsilon/v)^{1/2}N_o \tag{4}$$

In these equations D_o and D_s are the diameters of oil droplets and sediment grains, respectively, N_o is the oil droplet number concentration, $N_s(0)$ is the sediment number concentration at time zero, ε is the turbulent-kinetic-energy dissipation rate, ν is the kinematic viscosity of water, and α_{os} is the oil-sediment coalescence efficiency which was estimated to be of about 0.0016 from laboratory data of Payne et al. (1989).

More recently Khelifa et al. (2003b, 2004a,b) developed a Monte Carlo model for OMA formation (MCOMA). The model includes collision due to turbulence, differential settling and Brownian diffusion. It is adapted to consider both monodispersed and polydispersed sediment and droplet populations. Simulations showed that the time scale of OMA formation varies linearly (in logarithmic scale) with the percentage of OMA-stabilized oil if the sediment concentration is sufficient (Fig. 4). However, the time scale increases dramatically when the sediment concentration becomes critical to stabilize a given percentage of the reacting oil. The increase is higher with finer sediment (data with small size ratios in Figure 4). For instance, in the example shown in Figure 4, stabilization of about 20% of the initial oil requires a reacting time from less than a minute to about 17 h (over 10^3 min) when concentration ratio is 0.29.

3. Significance to oil spill remediation

Past oil spills have been a great *in situ* laboratory for scientists to gauge the significance of OMA formation to oil spill remediation. Among the lessons that remediation experts have learned from past spills are: (1) formation of OMA enhances biodegradation of the spilled oils, and (2) the observed efficacy of the *sediment relocation* technique (also called *surf-washing*) to cleanup oiled beaches is due to the enhancement of OMA formation (Lee et al., 1997, 2003a; Wood et al., 1998; Owens, 1999; Lee, 2002; Owens and Lee, 2003). Sediment relocation technique consists of transporting mechanically (using bulldozers for instance) oiled beach sediment from the upper tidal zone to the surf zone to enhance interaction with waves. Exposure to wave action enhances the mechanical abrasion, dispersion of oil into small droplets, and the rate of interaction between fine sediment particles and the dispersed oil.

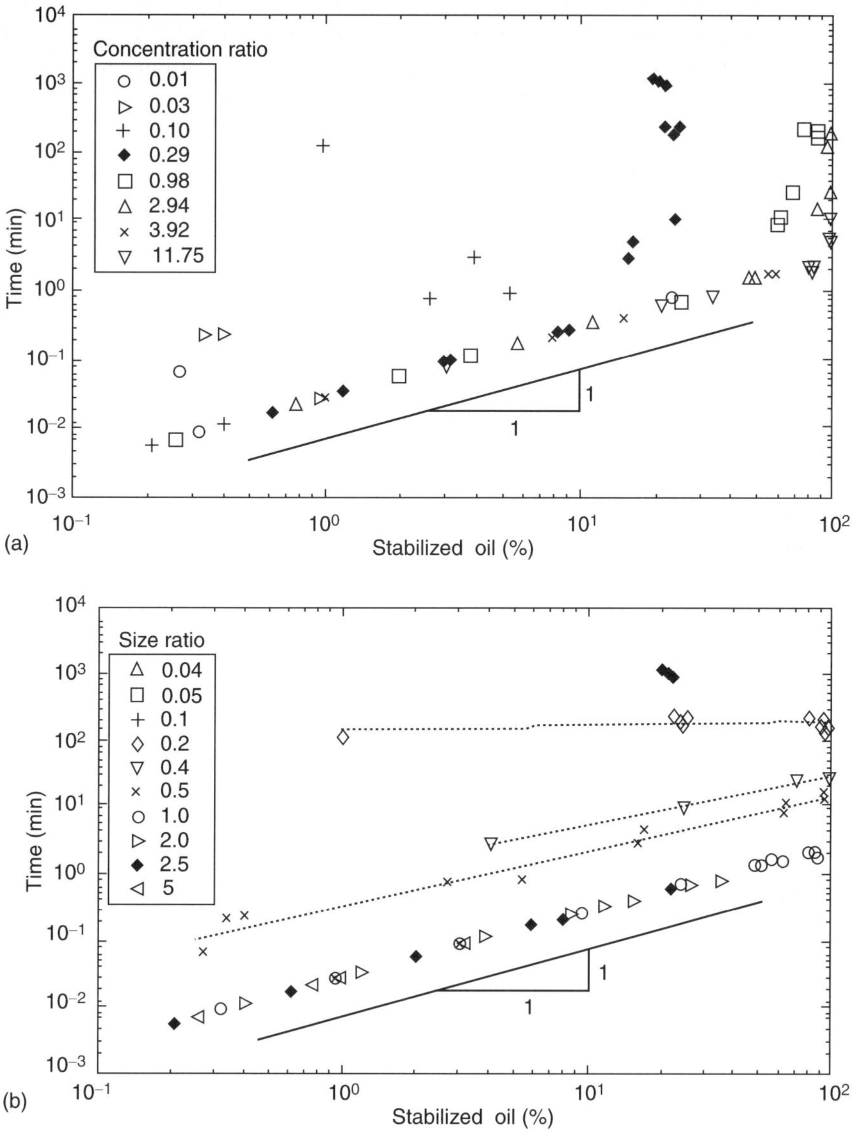

Figure 4. Variations of the formation time with the percentage of stabilized oil: (a) for different of sediment and droplet sizes (size ratio); (b) and different ratios of sediment and oil concentration (concentration ratio) concentration ratio. Results were obtained with light oil with $\rho_o = 850\ \mathrm{kg/m^3}$ (the dotted and solid lines were plotted to show trends only).

As reported by Lee (2002), the efficacy of sediment relocation to promote OMA induced oil dispersion has been demonstrated at two major oil spills (1993 Tampa Bay spill in Florida; 1996 Sea Empress spill in Wales, UK) and at a large-scale field experiment (1997 field trial in Svalbard, Norway). During the 1993 Tampa Bay response operations a beach had been manually cleaned of a no. 6 fuel oil. Then, bulldozers and front-end loaders were used to push the sediments into the surf zone to promote oil dispersion by OMA formation. The action was highly successful in removing oil stain from this high amenity beach comprised of fine to coarse sand with high shell fragment content and virtually no clays (Owens et al., 1995).

During response operations following the Sea Empress spill, scientists noted that oil along some shorelines was not adhering strongly to beach sediments (Colcomb et al., 1997). High turbidity in the water column suggested that OMA formation was a likely explanation. Subsequent microscopic observations showed that OMA formed readily in agitated mixtures of emulsified oil and water from the surf zone (Lee et al., 1997). A spill response strategy was developed to increase natural cleaning rates through dispersion of oil by OMA. This was accomplished by repeatedly relocating oiled sediments from the high-water mark into the intertidal zone over a 4-day period, to test the premise that OMA formation would be stimulated by wave-induced turbulence and high suspended sediment loads. This sediment relocation strategy proved successful. After 4 days of treatment, most of the oil emulsion was removed from cobbles, likely as OMA. It was estimated that approximately half of the oil released from the cobble was dispersed in association with mineral fines (Lunel et al., 1995, 1996). Gas chromatography–mass spectroscopy analysis confirmed that the treatment enhanced the biodegradation rates of the oil dispersed in association with mineral fines (Lee et al., 1997).

A large scale field experiment was carried out in Svalbard, Norway in 1997 to determine if sediment relocation could accelerate natural oil removal processes (Sergy et al., 2003; Guénette et al., 2003). Enhanced formation of OMA was observed following sediment relocation operations and gas chromatography–mass spectroscopy analysis demonstrated that a significant fraction of the oil dispersed into the nearshore waters and sediments in association with OMA was biodegraded (Lee et al., 2003b). Only limited concentrations of residual oil were found stranded on the shore in areas adjacent to the treated experimental plots and within nearshore subtidal sediments (Owens et al., 2003). This evidence suggested that the oil in association particles were reported to be relatively buoyant. With respect to potential secondary detrimental effects from the release of stranded oil from the beaches, none of the nearshore sediment samples exceeded the Canadian regulatory toxicity limit (Microtox_Test) for dredged spoils destined for ocean disposal (Lee et al., 2003b).

4. Further research required

From the oil spill remediation perspective, the main concern in studying OMA formation is to estimate, with acceptable accuracy, the amount of spilled oil that can be dispersed by this process under given spill and environmental conditions. Formation of OMA is not fully understood to do so, because effects of many factors are not known yet. Theoretical modeling is also at its early stage and necessitates further laboratory and field data for validation and development of behavior models. As a consequence, as far as the present authors know, none of the existing oil spill models includes OMA process in the simulations. Perhaps, this is because the process has been recognized recently only by the oil spill experts, and also due to the lack of behavior models to describe the process.

To improve our understanding of OMA formation and to move toward development of adequate behavior models, the following researches are suggested.

As mentioned previously, essentially two processes control the formation of OMA: formation of oil droplets from an oil slick and the aggregation between these oil droplets and suspended sediment grains or flocs. Formation of oil droplets is a key process not only for OMA formation, but also for modeling oil spill dispersion in general. Except the extensive laboratory work of Delvigne et al. (1987), very little has been done on droplet formation by researchers dealing with oil spills. For instance, the effects of oil weathering, temperature and oil composition on droplet-size distribution are still unknown. Research is needed to investigate effects of these parameters on formation of oil droplets and OMA.

In terms of heteroaggregation between oil droplets and sediment grains, research is still lacking on effects related to sediment phase, oil phase and environmental conditions. For sediment phase, prediction of effects related to sediment concentration, size and composition on OMA formation are still requiring further experimental investigations. Laboratory and field works are recommended to determine these effects. The ultimate goals from such research are (1) to establish the possible relationship between the OMA-stabilized oil concentration and sediment size and concentration. Results from this investigation will be compared with the data from Guyomarch et al. (1999) where the oil was chemically dispersed and with the data of Ajijolaiya (2004) obtained with Flotta oil (UK); (2) to determine the optimum range of sediment size for OMA formation. The findings will be compared with theoretical results obtained by Khelifa et al. (2004a); (3) to determine the MSC below which formation of OMA does not occur and how this MSC varies with sediment size. The findings will be compared with the recent simulations performed by Khelifa et al. (2004b); (4) to quantify the effects of sediment composition. This will validate the data obtained by Wood et al. (1998).

Oil type affects OMA formation through its effects on droplet formation from oil slicks or oil stranded on the shorelines. Nevertheless, a previous study (Khelifa et al., 2002) showed that asphaltenes-resins content enhances formation of OMA. It is expected that the presence of these components into oil droplets enhances their stickiness with sediment grains. It is also expected that sediment composition affects this stickiness as their hydrophobicity can change with the composition. It is recommended to conduct two series of laboratory experiments to show how sediment composition and asphaltenes-resins content affect the sticking efficiency between oil and sediment particles. Very little is known about this issue. Information from such study would be a valuable input to Monte Carlo modeling of OMA formation. Furthermore, experiment with different oils of different densities would validate the density effect shown in the previous numerical simulations (Khelifa et al. 2003b, 2004a,b).

Among environmental factors that have strong effects on OMA formation and are still poorly understood is mixing energy. This factor affects OMA formation through its effects on droplet formation, collision rate between oil and sediment particles, and growth of sediment flocs. The laboratory study conducted by Delvigne et al. (1987) has shown significant effects of mixing energy on droplet formation and empirical relationships were proposed. But, according to the authors, their data on OMA formation were altered by the experimental procedure they used. Experiments, performed, preferably in a wave tank, considering different mixing energy (wave conditions) and different oils are recommended to establish relationships which relate droplet-size distribution and OMA-stabilized oil concentration to mixing energy. Salinity is also another environmental factor that affects strongly OMA formation and its effects are not fully understood yet. Specifically, to establish variations of the critical aggregation salinity (parameter S_{cas} in Eq. 1) with oil and sediment types, further experimental research is essential.

Finally, laboratory data are required to validate existing theoretical models on the kinetics of OMA formation (Hill et al., 2002; Khelifa et al., 2003b). In a closely related issue, very little is known about the stability of OMA after their formation. From theoretical considerations, Levine et al. (1989) showed that an isolated solid sphere, partially wettable by both oil and water, is trapped at the oil/water interface in a very deep energy well. Laboratory tests can provide valuable information about the fate of OMA if the same samples are observed periodically for a relatively long period (a year or two) after their formation. Also, as mentioned by Muschenheim and Lee (2002), very little is known about the buoyancy in general and the settling velocities in particular of OMA. This information is important to understand the fate of OMA in aquatic environment. It is also required for the integration of OMA process into oil spill models.

5. Conclusion

Previous laboratory and field studies have shown that formation of OMA occurs naturally when sediment and oil are mixed artificially or naturally (due to wave action, for instance) in both seawater and freshwater. Several field investigations have confirmed the occurrence of this process at different oiled shorelines from past oil spills. The same studies showed that formation of OMA enhances also biodegradation of spilled oil. The efficacy of the sediment relocation technique to cleanup oiled beaches is due to enhancement of OMA formation.

Many laboratory studies have contributed to understanding the controlling parameters of OMA formation. It is now established that salinity, oil type, sediment size and concentration as well as temperature play major roles in OMA formation. Emergence of theoretical modeling of OMA formation in the last 2 years seems promising, as good agreements between predictions and observations were shown in related studies. Theoretical models are adequate tools to assess the significance of OMA formation to cleanup contaminated sites following accidental oil spills.

However, this study showed that further laboratory and field studies are required to improve the actual understanding of OMA formation. Further understanding of this natural process can lead, perhaps, to development of new techniques to enhance the significance of this process to cleanup oiled shorelines. Among the factors that are expected to play a significant role in OMA formation are the mixing energy, or more precisely wave conditions, and the weathering. These two factors have strong controls on droplet formation.

Acknowledgements

This study is part of a program to develop numerical models to simulate heteroaggregation of oil and sediment after an accidental oil spill. It is supported by the Petroleum Research

Atlantic Canada (previously Atlantic Canada Petroleum Institute) and the Natural Sciences and Engineering Research Council of Canada.

References

Ajijolaiya, L.O., 2004. The effects of mineral size and concentration on the formation of oil-mineral aggregates. Master Thesis, Department of Civil Engineering, Faculty of Engineering, Dalhousie University, Halifax, Canada, p. 80.

Bassin, N.J., Ichiye, T., 1977. Flocculation behavior of suspended sediments and oil emulsions. J. Sediment. Petrol. 47 (2), 671–677.

Binks, B.P., Lumsdon, S.O., 2000a. Catastrophic phase inversion of water-in-oil emulsions stabilized by hydrophobic silica. Langmuir 16, 2539–2547.

Binks, B.P., Lumsdon, S.O., 2000b. Influence of particle wettability on the type and stability of surfactant-free emulsions. Langmuir 16, 8622–8631.

Bragg, J.R., Owens, E.H., 1994. Clay–oil flocculation as a natural cleansing process following oil spills: part 1 – studies of shoreline sediments and residues from past spills, Proceedings of the 17th Arctic and Marine Oilspill Program (AMOP) Technical Seminar, Ottawa, Ontario, pp. 1–23.

Bragg, J.R., Yang, S.H., 1995. Clay–oil flocculation and its role in natural cleansing in Prince William Sound following the Exxon Valdez oil spill. In: Wells, P.G., Butler, J.N., Hughes, J.S. (Eds), Fate and Effects in Alaskan Waters, ASTM STP 1219. American Society for Testing and Materials, Philadelphia, PA, pp. 178–214.

Bragg, J.R., Prince, R.C., Harner, E.J., Atlas, R.M., 1994. Effectiveness of bioremediation for the Exxon Valdez oil spill. Nature 368 (6470), 413–418.

Colcomb, K., Bedborough, D., Lunel, T., Swannell, R., Wood, P., Rusin, J., Bailey, N., Halliwell, C., Davies, L., 1997. Shoreline cleanup and waste disposal issues during the Sea Empress Incident, Proceedings of the 1997 International Oil Spill Conference. American Petroleum Institute, Washington, DC, pp. 195–203.

Delvigne, G.A.L., 1987. Droplet size distribution of naturally dispersed oil. In: Kuiper, J., Van den Brink, W.J. (Eds), Fate and effects of oil in marine ecosystems. Martinus Nijhoff Publications, Dordrecht, The Netherlands, pp. 29–40.

Delvigne, G.A, van der Stel, J.A., Sweeny, C.E., 1987. Measurement of vertical turbulent dispersion and diffusion of oil droplets and oiled particles. Final report: OCS Study MMS 87-111. US Department of the Interior, Minerals Management Service, Anchorage, Alaska, p. 501.

Guénette, C.C., Sergy, G.A., Owens, E.H., Prince, R.C., Lee, K., 2003. Experimental design of the Svalbard shoreline field trials. Spill Sci. Technol. Bull. 8 (3), 245–256.

Guyomarch, J., Merlin, F., Bernanose, P., 1999. Oil interaction with mineral fines and chemical dispersion: behaviour of the dispersed oil in coastal or estuarine conditions, Environment Canada's 22nd Artic and Marine Oilspill (AMOP) Technical Seminar, Calgary, Alberta, Canada, pp. 137–149.

Guyomarch, J., Le Floch, S., Merlin, F., 2002. Effect of suspended mineral, water salinity and oil type on the size of oil–mineral aggregates in the presence of chemical dispersant load. Spill Sci. Technol. Bull. 8 (1), 95–100.

Hill, P.S., Khelifa, A., Lee, K., 2002. Time scale for oil droplet stabilization by mineral particles in turbulent suspensions. Spill Sci. Technol. Bull. 8 (1), 73–81.

Hoffman, R.W., Shier, L., 1999. Additional investigations of oil-sediment interactions in Alaskan North Slope and TAP area Rivers, Environment Canada's 22nd Arctic and Marine Oilspill (AMOP) Technical Seminar, Calgary, Alberta, Canada.

Huang, C.P., Elliot, H.A., 1977. The stability of emulsified crude oils as affected by suspended particles. In: Wolfe, D.A. (Ed.), Fate and Effect of Petroleum Hydrocarbons on Marine Organisms and Ecosystems. Pergamon Press, New York, pp. 413–420.

Jézéquel, R., Merlin, F.-X., Lee, K., 1999. The influence of microorganisms on oil–mineral fine interactions in low energy coastal environment, Proceedings of the 1999 International Oil Spill Conference. American Petroleum Institute, Washington, DC, Publication no. 4686B, pp. 771–775.

Kepkay, P., Bugden, J.B.C., Lee, K., Stoffyn-Egli, P., 2002. Application of ultraviolet fluorescence (UVF) spectroscopy to monitor oil–mineral aggregate (OMA) formation. Spill Sci. Technol. Bull. 8 (1), 101–108.

Khelifa, A., Stoffyn-Egli, P., Hill, P.S., Lee, K., 2002. Characteristics of oil droplets stabilized by mineral particles: the effect of oil types and temperature. Spill Sci. Technol. Bull. 8 (1), 19–30.

Khelifa, A., Stoffyn-Egli, P., Hill, P.S., Lee, K., 2003a. Characteristics of oil droplets stabilized by mineral particles: the effect of salinity, 2003 International OilSpill Conference, Vancouver, BC, Canada, pp. 1–8.

Khelifa, A., Hill, P.S., Lee, K., 2003b. A stochastic model to predict the formation of oil-mineral aggregates, 26th Proceedings of Arctic and Marine OilSpill Program (AMOP) Technical Seminar, Vol. 2. Environment Canada, Victoria, BC, Canada, pp. 893–909.

Khelifa, A., Hill, P.S., Lee, K., 2004. Modelling the effect of sediment size on OMA formation. 27th Proceedings of Arctic and Marine OilSpill Program (AMOP) Technical Seminar, Environment Canada, Edmonton, Alta., Canada.

Khelifa, A., Hill, P.S., Lee, K., 2004. The role of oil-sediment aggregation in dispersion and biodegradation of spilled oil. This volume.

Khelifa, A., Hill, P.S., Stoffyn-Egli, P., Lee, K. 2005. Effects of salinity and clay composition on oil–clay aggregations. Mar. Environ. Res. 59, 235–254.

Lee, K., 2002. Oil–particle interactions in aquatic environments: influence on the transport, fate, effect and remediation of oil spills. Spill Sci. Technol. Bull. 8 (1), 3–8.

Lee, K., Stoffyn-Egli, P., 2001. Characterization of oil-mineral aggregates, Proceedings of the 2001 International Oil Spill Conference. American Petroleum Institute, Washington, DC, pp. 991–996.

Lee, K., Lunel, T., Wood, P., Swannell, R., Stoffyn-Egli, P., 1997. Shoreline cleanup by acceleration of clay–oil flocculation processes, Proceedings of the 1997 International Oil Spill Conference. American Petroleum Institute, Washington, DC, Publication no. 4651, pp. 235–240.

Lee, K., Stoffyn-Egli, P., Wood, P.A., Lunel, T. 1998. Formation and structure of oil-mineral fines aggregates in coastal environments. Proceedings of the 21st Arctic and Marine Oilspill Program (AMOP) Technical Seminar. Environment Canada, Ottawa, Ont., pp. 911–921.

Lee, K., Stoffyn-Egli, P., Owens, E.H., 2002. The OSSA II pipeline oil spill: natural mitigation of a riverine oil spill by oil–mineral aggregate formation. Spill Sci. Technol. Bull. 7 (3–4), 149–154.

Lee, K., Stoffyn-Egli, P., Tremblay, G.H., Owens, E.H., Sergy, G.A., Guénette, C.C., Prince, R.C., 2003a. Oil-mineral aggregate formation on oiled beaches: natural attenuation and sediment relocation. Spill Sci. Technol. Bull. 8 (3), 285–296.

Lee, K., Wohlgeschaffen, G., Tremblay, G.H., Johnson, B., Sergy, G.A., Prince, R.C., Guenette, C.C., Owens, E.H., 2003b. Toxicity evaluation with the Microtox test to assess the impact of in-situ oiled shoreline treatment options: natural attenuation and sediment relocation. Spill Sci. Technol. Bull. 8 (3), 273–284.

Le Floch, S., Guyomarch, J., Merlin, F.X., Stoffyn-Egli, P., Dixon, J., Lee, K., 2002. The influence of salinity on oil-mineral aggregate formation. Spill Sci. Technol. Bull. 8 (1), 65–71.

Levine, S., Bowen, B.D., Partridge, S.J., 1989. Stabilization of emulsions by fine particles I. Partitioning of particles between continuous phase and oil/water interface. Colloid. Surf. 38, 325–343.

Lunel, T., Swannell, R.P., Rusin, J., Wood, P., Bailey, N., Halliwell, C., Davies, L., Sommerville, M., Dobie, A., Mitchell, D., McDonagh, M., Lee, K., 1995. Monitoring the effectiveness of response operations during the Sea Empress incident: a key component of the successful counter-pollution response. Spill Sci. Technol. Bull. 2, 99–112.

Lunel, T., Lee, K., Swannell, R., Wood, P., Rusin, J., Bailey, N., Halliwell, C., Davies, L., Sommerville, M., 1996. Shoreline cleanup during the Sea Empress incident: the role of surf washing (clay–oil flocculation), dispersants and bioremediation, Proceedings of the 19th Arctic and Marine Oilspill Program (AMOP) Technical Seminar. Environment Canada, Ottawa, Ontario, pp. 1521–1540.

McCourt, J., Shier, L., 1998. Interaction between oil and suspended particulate matter in the Yukon River, Environment Canada's 21st Arctic and Marine OilSpill Program (AMOP) Technical Seminar, Ottawa, Ont., Canada, pp. 77–88.

McCourt, J., Shier, L., 2000. Preliminary findings of solids interaction in eight Alskan Rivers, Environment Canada's 23rd Arctic and Marine Oilspill Program (AMOP) Technical Seminar. Vancouver, British Columbia, Canada.

Menon, V.B., Wasan, D.T., 1988. Characterization of oil–water interfaces containing finely divided solids with applications to the coalescence of water-in-oil emulsions: a review. Colloid. Surf. 29, 7–27.

Meyers, P.A., Quinn, J.G., 1973. Association of hydrocarbons and mineral particles in saline solution. Nature 244, 23–24.

Muschenheim, D.K., Lee, K., 2002. Removal of oil from the sea surface through particulate interactions: review and prospectus. Spill Sci. Technol. Bull. 8 (1), 9–18.

Omotoso, O.E., Munoz, V.A., Mikula, R.J., 2002. Mechanisms of crude oil–mineral interactions. Spill Sci. Technol. Bull. 8 (1), 45–54.

Owens, E.H., 1999. The interaction of fine particles with stranded oil. Pure Appl. Chem. 71 (1), 83–93.

Owens, E.H., Lee, K., 2003. Interaction of oil and mineral fines on shorelines: review and assessment. Mar. Pollut. Bull. 47 (9–12), 397–405.

Owens, E.H., Bragg, J.R., Humphrey, B., 1994. Clay–oil flocculation as a natural cleansing process following oil spills: part 2 – Implications of study results in understanding past spills and for future response decisions, Proceedings 17th Arctic and Marine Oil Spill Program (AMOP) Technical Seminar, Edmonton, Alberta, pp. 25–37.

Owens, E.H., Davis, R.A. Jr., Michel, J., Stritzke, K., 1995. Beach cleaning and the role of technical support in the 1993 Tampa Bay spill, Proceedings of the 1995 International Oil Spill Conference. American Petroleum Institute, Washington, DC, pp. 627–634.

Owens, E.H., Sergy, G.A., Guenette, C.C., Prince, R.C., Lee, K., 2003. The reduction of stranded oil by in-situ shoreline treatment options. Spill Sci. Technol. Bull. 8 (2), 257–272.

Payne, J.R., Clayton, G.D., Jr., McNabb, C.L., Kirstein, C.L., Clary, R.T., Redding, J.S., Evans, E., Reimnitz, Kempema, E.W., 1989. Oil-ice-Sediment Interactions During Freezeup and Breakup. US Department of Commerce, NOAA, OCSEAP. Final Report 64, p. 382.

Payne, J.R., Clayton, J.R. Jr., Kirstein, B.E., 2003. Oil/suspended particulate material interactions and sedimentation. Spill Sci. Technol. Bull. 8 (2), 201–221.

Pickering, S.U., 1907. J. Chem. Soc. 91.

Poirier, O.A., Thiel, G.A., 1941. Deposition of free oil by sediments settling in sea water. Bull. Am. Assoc. Petrol. Geol. 25 (12), 2170–2180.

Ramsden, W., 1903. Proc. R. Soc. London, 72.

Sergy, G., Guennette, C., Owens, E., Prince, R., Lee, K., 1999. Treatment of oiled sediment shorelines by sediment relocation, Proceedings of the 1999 International Oil Spill Conference, Seattle, Washington, USA, pp. 549–554.

Sergy, G.A., Guenette, C.C., Prince, R.C., Lee, K., 2003. In-situ treatment of oiled sediment Shorelines. Spill Sci. Technol. Bull. 8 (3), 237–244.

Stoffyn-Egli, P., Lee, K., 2002. Formation and characterization of oil-mineral aggregates. Spill Sci. Technol. Bull. 1, 31–44.

Weise, A.M., Nalewajko, C., Lee, K., 1999. Oil-mineral fine interactions facilitate oil biodegradation in seawater. Environ. Technol. 20, 811–824.

Wood, P.A., Lunel, T., Daniel, F., Swannell, R., Lee, K., Stoffyn-Egli, P., 1998. Influence of oil and mineral characteristics on oil-mineral interaction. Proceedings of the 21st Arctic and Marine Oilspill Program (AMOP) Technical Seminar, Environment Canada, Ottawa, Ont., pp. 51–77.

Yan, N., Masliyah, J.H., 1995. Characterization and demulsification of solids-stabilized oil-in-water emulsions. Part 1. Partitioning of clay particles and preparation of emulsions. Colloid. Surf. 96, 229–242.

Zurcher, F., Thuer, M., 1978. Rapid weathering processes of fuel oil in natural waters: analyses and Interpretations. Environ. Sci. Technol. 12 (7), 838–843.

Oil Pollution and its Environmental Impact in the Arabian Gulf Region
M. Al-Azab, W. El-Shorbagy and S. Al-Ghais, editors

Chapter 11

Modelling seawater–pore water exchange near seabeds

Arzhang Khalili

Max Planck Institute for Marine Microbiology, Celsiusstr.1, 28359 Bremen, Germany

Abstract

When sediments are permeable, a direct link between the pore water and overlying water column is given. In such cases – compared to impermeable sediments – the transport of nutrients and chemicals from the seawater into the seabed (and vice versa) is enhanced. So far, the conventional approach to study the interfacial phenomena has been to collect sediment samples and study them in the lab. The purpose of this chapter is to present a mathematical method that is able to quantify interfacial fluxes near and through preamble interfaces, simultaneously, without using any hydrodynamic interface conditions. The results of this model have been verified in non-invasive experiments. The model may be applied for calculation of contaminant transport through permeable sediments.

1. Introduction and governing equations in a porous layer

The transport of nutrients and particulate matter across permeable sediment–water interfaces plays an important role in many situations of marine biogeochemistry such as in rivers, oceans and lakes. The quantification of the amount of substances transported through the interface is, however, a difficult task. Common experimental techniques for *in-situ* quantification of the interfacial exchange are micro sensors (Glud et al., 1995, 1996), benthic chambers (Gust, 1990), sediment cut-off (Huettel and Gust, 1992) or landers (Witte et al., 2003).

In many of these measurement methods, however, the real situation is perturbed due to the existence of devices intruded into the fluid or the porous bed, and additional unwanted flows are generated that may change the concentration field of the substance in question. Besides, the interfacial transport is governed by many different parameters such as the topography shape, the physical properties of the porous medium (permeability κ and porosity φ), and the velocity field. Therefore, an efficient mathematical representation for the interfacial exchange is a highly desired tool. The first mathematical attempt in this direction was the Darcy's law (1856) to account for the interaction of the flow density ($\mathbf{v}$) with the pressure gradient (∇p) given in modern vector notation as

$$\nabla p = -\frac{\mu}{\kappa} v \tag{1}$$

E-mail address: akhalili@mpi-bremen.de (A. Khalili).

with κ being the permeability of the porous medium and μ the fluid density. However, it is well-known tat the linear Darcy law holds for flows at low Reynolds numbers in which the driving forces are small and balanced by the viscous forces, solely. The Darcy's law is also found to be valid for homogeneous isotropic and non-deformable porous media. When these assumptions do not hold, the applicability of the Darcy's law is limited, and extensions need to be made to the linear Darcy equation.

First non-linear generalization of Darcy's law has been suggested by Forchheimer (1901), which is known as Darcy–Forchheimer equation given by

$$\nabla p = -\frac{\mu}{\kappa}\boldsymbol{v} - \frac{c_{\mathrm{f}}\rho_{\mathrm{f}}}{\sqrt{\kappa}}|\boldsymbol{v}|\boldsymbol{v} \tag{2}$$

with c_{f} and ρ_{f} denoting the form drag coefficient and the fluid density, respectively. The second extension to the Darcy's law was made by Brinkman (1947), and was intended to account for the viscous drag by adding $\tilde{\mu}\nabla^2\boldsymbol{v}$ ($\tilde{\mu}$ being the effective viscosity) to the Darcy equation. With this term included in the Darcy equation, the Brinkman-extended Darcy equation can be written by

$$\nabla p = \tilde{\mu}\nabla^2\boldsymbol{v} - \frac{\mu}{\kappa}\boldsymbol{v}. \tag{3}$$

Considering both extensions simultaneously results in

$$\nabla p = \tilde{\mu}\nabla^2\boldsymbol{v} - \frac{\mu}{\kappa}\boldsymbol{v} - \frac{c_{\mathrm{f}}\rho_{\mathrm{f}}}{\sqrt{\kappa}}|\boldsymbol{v}|\boldsymbol{v}. \tag{4}$$

Finally, including the total acceleration following Wooding (1957), one obtains the most general form of the momentum equation for a saturated homogeneous porous medium as

$$\rho_{\mathrm{f}}[\partial_t \boldsymbol{V} + (\boldsymbol{V}\cdot\nabla)\boldsymbol{V}] = -\nabla p + \tilde{\mu}\nabla^2\boldsymbol{v} - \frac{\mu}{\kappa}\boldsymbol{v} - \frac{c_{\mathrm{f}}\rho_{\mathrm{f}}}{\sqrt{\kappa}}|\boldsymbol{v}|\boldsymbol{v} \tag{5}$$

with ∂_t denoting differentiation with respect to time and $\cdot$ being the scalar product symbol. It should be noted that the intrinsic velocity $\mathbf{V}$ (average of velocity over a volume that contain fluid only) and the Darcy velocity $\mathbf{v}$ (average of velocity over a volume that contain both fluid and solid matrix) are linked to each other by the so-called Dupuit–Forchheimer relation (Kaviany, 1999) given as

$$\boldsymbol{v} = \varphi\boldsymbol{V} \tag{6}$$

With this equation, the general momentum equation would read as

$$\rho_{\mathrm{f}}[\varphi^{-1}\partial_t\boldsymbol{v} + \varphi^{-2}(\boldsymbol{v}\cdot\nabla)\boldsymbol{v}] = -\nabla p + \varphi^{-1}\tilde{\mu}\nabla^2\boldsymbol{v} - \frac{\mu}{\kappa}\boldsymbol{v} - \frac{c_{\mathrm{f}}\rho_{\mathrm{f}}}{\sqrt{\kappa}}|\boldsymbol{v}|\boldsymbol{v}. \tag{7}$$

After non-dimensionalization with the characteristic length L, velocity U and time L/U, the flow model equations (conservation of momentum and mass) containing all extensions to Darcy's model in vector form are (see for example Prasad, 1991):

$$[\partial_t\boldsymbol{v} + \varphi^{-1}(\boldsymbol{v}\cdot\nabla)\boldsymbol{v}] = -\varphi\nabla p + \Lambda\frac{\varphi}{\mathrm{Re}}\nabla^2 - B\varphi\left(\frac{1}{\mathrm{Re\ Da}}\boldsymbol{v} + \mathrm{F}|\boldsymbol{v}|\boldsymbol{v}\right), \tag{8}$$

$$\nabla\cdot\boldsymbol{v} = 0. \tag{9}$$

The parameters appearing in Eq. (8), namely Λ, Re, Da and F are, respectively, the viscosity ratio, Reynolds number, Darcy number, and the Forchheimer number and are given by

$$\Lambda = \frac{\tilde{\mu}}{\mu}, \tag{10}$$

$$\mathrm{Re} = \frac{LU}{\upsilon}, \tag{11}$$

$$\mathrm{Da} = \frac{\kappa}{L^2}, \tag{12}$$

$$\mathrm{F} = \frac{c_\mathrm{f}}{\sqrt{\mathrm{Da}}}. \tag{13}$$

The parameter B is a binary constant that may take the values zero or unity, and will be explained in the next section.

Likewise, the concentration (c) equation in dimensionless form can be given as

$$\partial_\mathrm{t} c + \varphi^{-1}(\mathbf{v}\cdot\nabla)c = \frac{\Gamma}{\mathrm{Re\ Sc}}\nabla^2 c \tag{14}$$

where Γ and Sc are the diffusivity ratio and the Schmidt number given by

$$\Gamma = \frac{\tilde{\sigma}}{\sigma}, \tag{15}$$

$$\mathrm{Sc} = \frac{\upsilon}{\sigma}, \tag{16}$$

with $\tilde{\sigma}$ and σ denoting the diffusivities of c inside the porous and fluid layer.

2. Boundary and interface condition and equations in composite fluid–sediment layer

While the boundary conditions at free fluid surfaces and solid walls can be formulated in a straight forward manner, a clear consensus on the mathematical form of the hydrodynamic condition that prevail at the interface between a fluid and a sediment layer, does not exist.

Beavers and Joseph (1967) were the first who formulated a slip condition, later extended further by Saffman (1971). Since then, the Beavers and Joseph condition has opened a vital discussion on the interface condition problem. There are other suggestions that include the continuity of velocity, pressure and shear across the interface (Taylor, 1971; Ross, 1983; Vafai and Kim, 1990) or, more recently, the stress-jump condition of Ochoa-Tapia and Whitaker (1995).

It should be noted that all these suggestions have been derived for one-dimensional problems, and that the formulation of the correct set of interface conditions for three-dimensional flows is seldom obvious.

However, it is possible to solve the momentum equation (Eq. 8) in the vicinity of an interface without applying any interface condition. By proper choice of parameter B, φ, Λ and Γ in Eq. (8), namely, the momentum equation in a porous layer as well as fluid region may be obtained when setting:

$$B = \begin{cases} 0, & \text{in fluid region} \\ 1, & \text{in porous region} \end{cases}$$

$$\varphi = \begin{cases} 1, & \text{in fluid region} \\ 0 < \varphi < 1, & \text{in porous region} \end{cases}$$

$$\Lambda = \begin{cases} 1, & \text{in fluid region} \\ \neq 1, & \text{in porous region} \end{cases}$$

and

$$\Gamma = \begin{cases} 1, & \text{in fluid region} \\ \neq 1, & \text{in porous region} \end{cases}$$

Hence, Eqs (8), (9) and (14) may be treated as equations of a single domain in which different input parameters are taken, and therefore matching of variable values across the interface is inherent in the formulation itself, thus doing away with the need for separate interface conditions.

Further details are given by Basu and Khalili (1999).

3. Solution method

The Eqs (8), (9), and (14) need to be put into a proper coordinate system depending on the geometry of the problem in question. Using approximation methods such as finite differences, finite elements or finite volumes, discretized form of the above equations may be obtained. The resulting systems of algebraic equations can then be solved by standard numerical techniques. An effective solution method based on the MAC algorithm combined with the fractional step method (Kim and Moin, 1985) is given by Khalili et al. (1997).

4. Application and examples

Figure 1 demonstrates the field of application of the mathematical method developed. As can be seen it contains of the flow and concentration distribution over impermeable as well as permeable beds.

The flow may be driven by simple currents, waves or a combination of both. The effect of the activity of the living organisms, referred to as bio-irrigation, also may be included using the so-called irrigation parameter (Emerson et al., 1984; Aller, 1990).

In order to verify the results of the method developed, first a cylindrical geometry with a rotating lid was constructed that was filled up to half height with a permeable medium and up to top with water. When the disk was set into rotation, a three-dimensional motion was generated that transported fluid, and with this, substances into the sediment layer underneath (see Fig. 2).

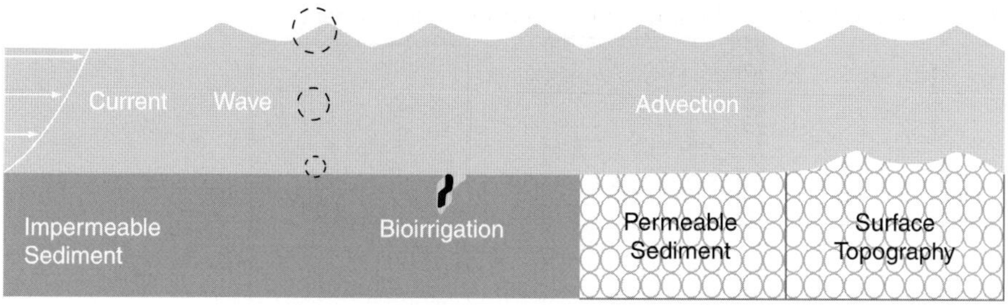

Figure 1. Different possible applications include effect of currents and waves on impermeable sediments and effect of advection on permeable seabeds.

The flow and concentration field generated by the rotation of the disk was calculated (Basu and Khalili, 1999) and then compared with experimental results (Khalili et al., 1999) using different non-invasive techniques with good agreement between the simulations and the experiments as shown in Figure 3.

Another situation is given when seabeds with periodic ripples are subject to horizontal oscillatory motion. The seabed may be solid (Fig. 4), porous but impermeable (Fig. 7) or permeable (Fig. 8).

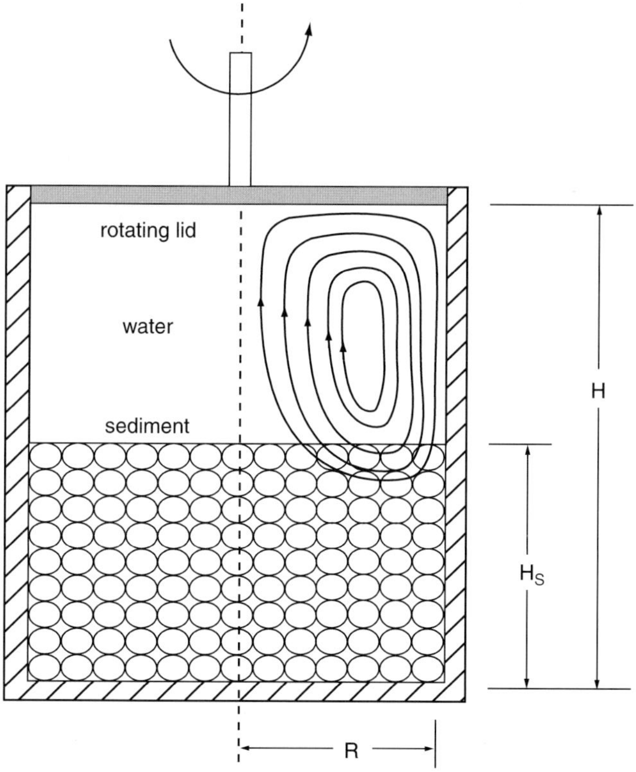

Figure 2. Interfacial exchange of fluid due to the secondary flow induced by the disk rotation.

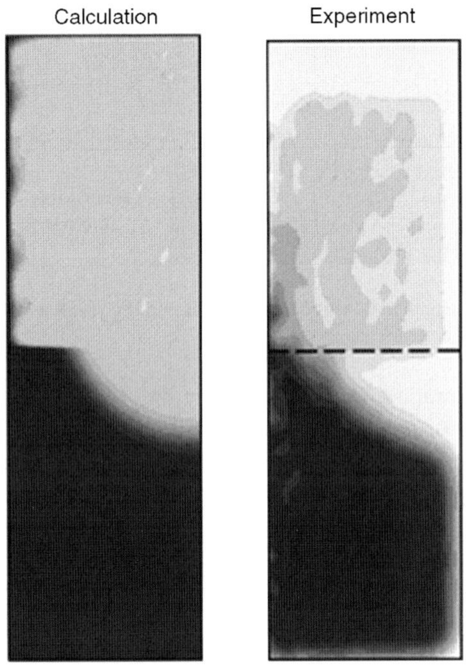

Figure 3. Comparison of the concentration profile in the cylindrical container by numerical simulation (left image) and experiments (right image) with PET (only half of the container is shown here).

In the first case (solid ripples), the computational domain contains only fluid, and the flow equations are given by the Navier–Stokes equations.

A standing wave given by an oscillatory function drives the flow over solid ripples of known wave number. The complete flow cycle is shown by the stream function distribution within one period (Fig. 5).

As can be seen therein, the flow is confined to a narrow layer in the vicinity of the solid bed. The corresponding concentration distribution may be seen from Figure 6.

An interesting feature of the program is demonstrated by Figures 7 and 8, where an impermeable seabed (Fig. 7) or permeable one (Fig. 8) replaced the solid bed of Figure 4. These programs developed may easily switch to any case desired. Figure 7 demonstrates the concentration distribution over an impermeable bed above the crest (left image) and

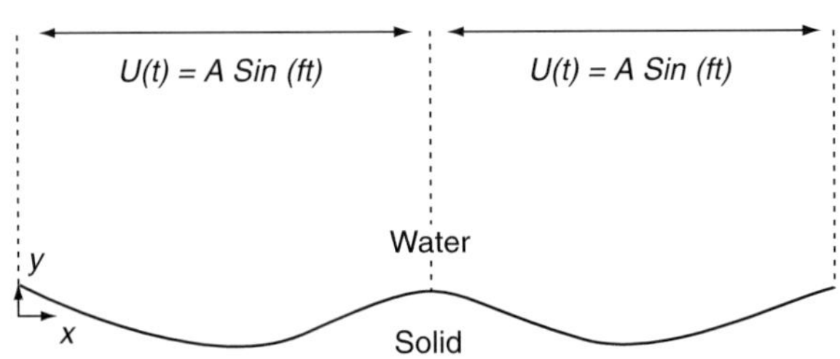

Figure 4. A sinusoidal oscillation over a periodic seabed topography.

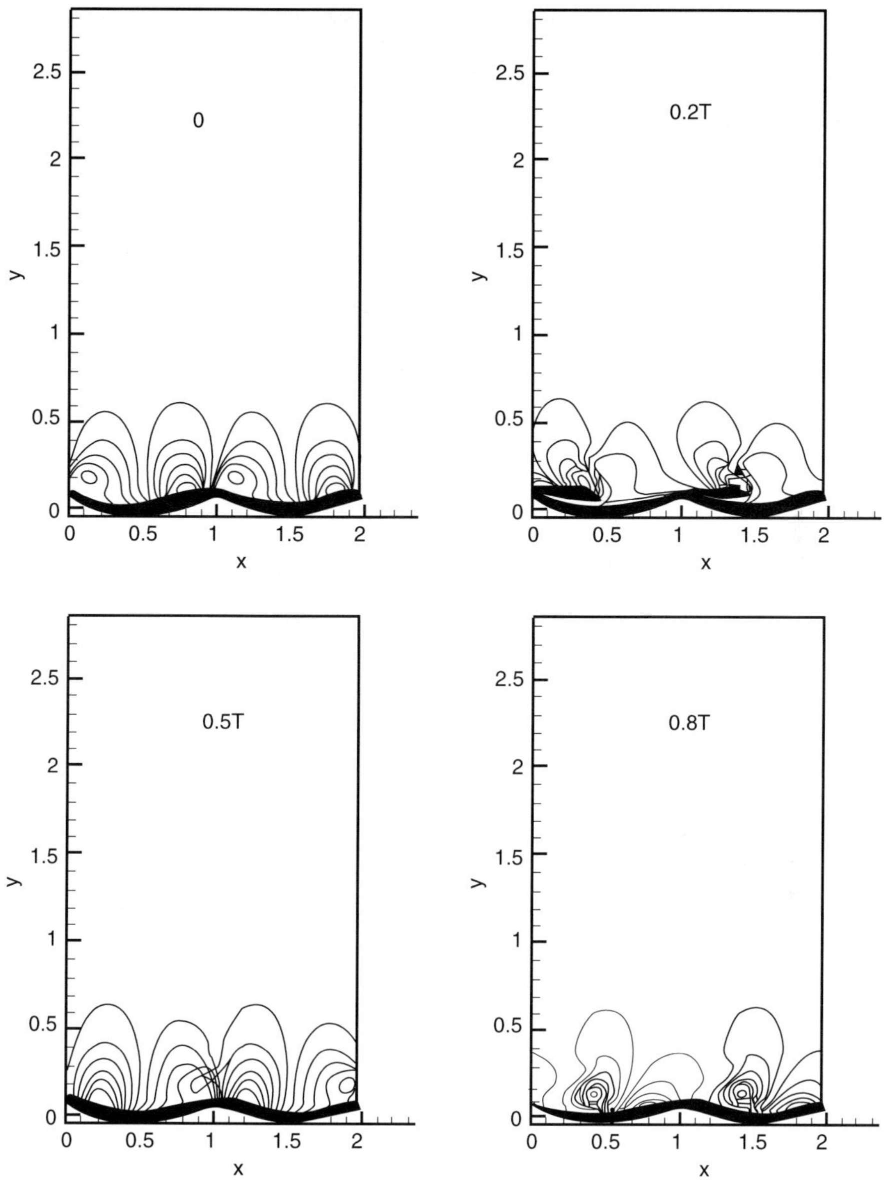

Figure 5. Stream function over solid ripples within one period.

the trough (right image). It is interesting to observe how the concentration of a given substance in the overlying water is transported into the impermeable bed underneath.

Finally, when a permeable bed is considered, the situation changes thoroughly, because of the advection of flow through the bed. In this situation, namely, a pressure gradient is formed which is responsible for the transport of substances dissolved in the fluid. This is shown in Figure 8 by the vorticity transport (left image). The corresponding concentration

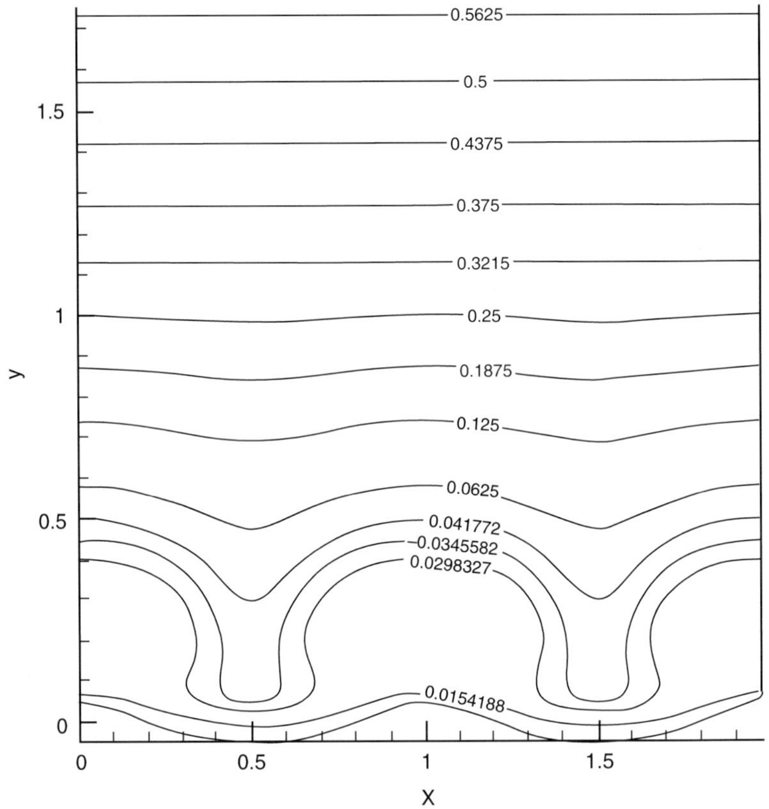

Figure 6. Concentration distribution over solid ripples.

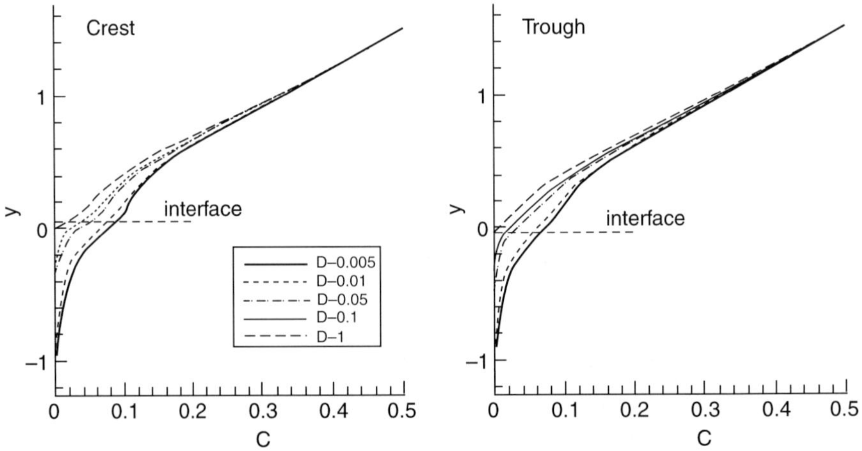

Figure 7. Concentration distribution over solid ripples (impermeable sediment).

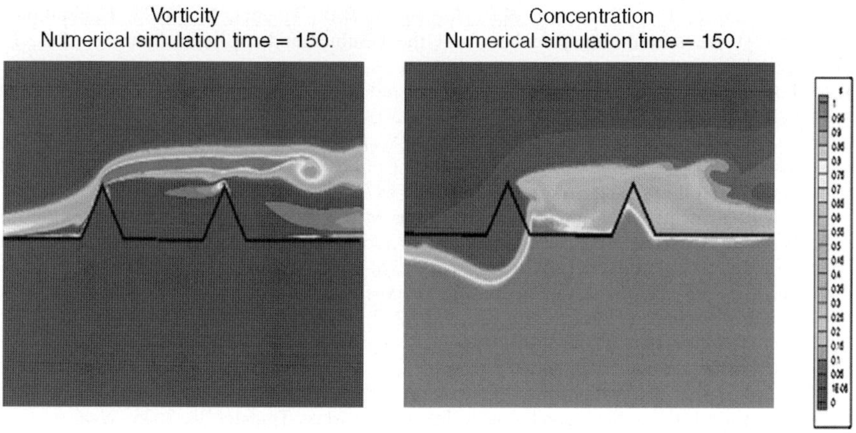

Figure 8. Vorticity transport (left image) and concentration distribution (right image) over a topography with two permeable ripples.

equation, demonstrating the penetration depth of the substances into the pore water is shown in the right image of Figure 8.

5. Conclusions

A numerical model and laboratory experiments were developed to study the flow field and concentration exchange at permeable water-sediment layers. With the help of such a method, percolation of pollutions into permeable seabeds may be traced and quantified. The application of such techniques may be extended to many other situations in science and nature where composite fluid–sediment layer occurs. It has been shown that the model developed (Khalili et al., 1997; Basu and Khalili, 1999) is capable of providing reliable results on exchange of matter and fluid over solid as well as porous beds with very low (almost impermeable) and high permeability.

References

Aller, R.C., 1990. Bioturbation and manganese cycling in hemipelagic sediments. Philos. Trans. R. Soc. London, 51–68.
Basu, A.J., Khalili, A., 1999. Computation of flow through a fluid–sediment interface in a benthic chamber. Phys. Fluids 11, 1395–1405.
Beavers, G.S., Joseph, D.D., 1967. Boundary conditions at a naturally permeable wall. J. Fluid Mech. 30, 197–207.
Brinkman, H.C., 1947. A calculation of the viscous force exerted by a flowing fluid on a dense swarm of particles. Appl. Sci. Res. A 1, 27–34.
Darcy, H., 1856. Les Fontaines Publiques de la Ville de Dijon. Dalmont, Paris.
Emerson, S., Jahnke, R., Heggie, D., 1984. Sediment-water exchange in shallow water estuarine sediments. J. Mar. Res. 42, 709–730.
Forchheimer, P., 1901. Wasserbewegung durch Boden. Z. Ver. Deutsch. Ing. 45, 1782–1788.

Glud, R.N., Gundersen, J.K., Revsbech, N.P., Jorgensen, B.B., Huettel, M., 1995. Calibration and performance of the stirred flux chamber from the benthic lander Elinor. Deep-Sea Res. I 42, 1029–1042.

Glud, R.N., Forster, S., Huettel, M., 1996. Influence of radial pressure gradients on solute exchange in stirred benthic chambers. Mar. Ecol. Prog. Ser. 141, 303–311.

Gust, G., 1990. Method of Generating Precisely Defined Wall Shearing Stresses, U.S. Patent, 4, 973,165.

Huettel, M., Gust, G., 1992. Solute release mechanisms from confined sediment cores in stirred benthic chambers and flume flows. Mar. Ecol. Prog. Ser. 82, 187–197.

Kaviany, M., 1999. Principles of Heat Transfer in Porous Media, 2nd ed. Springer, New York.

Khalili, A., Basu, A.J., Mathew, J., 1997. A non-Darcy model for recirculating flow a fluid–sediment interface in a cylindrical container. Acta Mech. 123, 75–87.

Khalili, A., Basu, J., Pietrzyk, U., Raffel, M., 1999. An experimental study of recirculating flow through fluid–sediment interfaces. J. Fluid Mech. 383, 229–247.

Kim, J., Moin, P., 1985. Application of a fractional-step method to incompressible Navier–Stokes equations. J. Comput. Phys. 59, 308–323.

Ochoa-Tapia, J., Whitaker, S., 1995. Momentum transfer at the boundary between a porous medium and a homogeneous fluid I: theoretical development. Int. J. Heat Mass Transfer 38, 2635–2646.

Prasad, V., 1991. Convective heat and mass transfer in porous media. In: Kakaç, S. (Ed.), Convective Flow Interaction and Heat Transfer between Fluid and Porous Layer, NATO ASI Series E: Applied Science, Vol. 196, pp. 563–615.

Ross, S.M., 1983. Theoretical model of the boundary condition at a fluid–porous interface. AIChE J. 29, 840–846.

Saffman, P.G., 1971. On the boundary condition at the surface of a porous medium. Stud. Appl. Math. 50, 93–101.

Taylor, G.I., 1971. A model for the boundary condition of a porous material. Part 1. J. Fluid Mech. 49, 319–326.

Vafai, K., Kim, S., 1990. Fluid mechanics of the interface region between a porous medium and a fluid layer – an exact solution. Int. J. Heat Fluid Flow 11, 254–256.

Witte, U., Wenzhöfer, F., Sommer, S., Boetius, A., Heinz, P., Aberle, A., Sand, M., Cremer, A., Abraham, W.R., Jørgensen, B.B., Pfannkuche, O., 2003. In situ experimental evidence of the fate of a phytodetritus pulse at the abyssal sea floor. Nature 424, 763–766.

Wooding, R.A., 1957. Steady state free thermal convection of liquid in a saturated permeable medium. J. Fluid Mech. 2, 273–285.

IV

Environmental effects of oil pollution on the ecosystem components

Oil Pollution and its Environmental Impact in the Arabian Gulf Region
M. Al-Azab, W. El-Shorbagy and S. Al-Ghais, editors

Chapter 12

Effects of petroleum hydrocarbons on aquatic animals

M.A. Quddus Khan[1,*], Saif M. Al-Ghais[2], Bogdan Catalin[1] and Yaseen H. Khan[1]

[1]*Department of Biological Sciences, University of Illinois at Chicago, Chicago, IL 60607, USA*
[2]*Biology Department, Faculty of Science, UAE University, P.O. Box 17551, Al-Ain, United Arab Emirates*

1. Petroleum hydrocarbons (PH) in aquatic environment

Recent estimates of petroleum hydrocarbon (PH) concentrations in the Arabian Gulf average to about 26 ug/l in surface water (Douabul et al., 1987; Douabul and Al-Shiwafi, 1998) and about 5 ug/kg in bottom sediments (Aboul-Dahab and Almadia, 1993; Shriadah, 2000, 2001). About one-third of petroleum hydrocarbons are aromatic including four- and higher-ringed members (polycyclic aromatic hydrocarbons, PAH). With continued oil production and transport the concentrations of these hydrocarbons in water and sediment, at least in the highly trafficked waters, will keep increasing. Among aquatic organisms the oviparous animals may accumulate, by direct exposure or via trophic transport, high levels of these lipophilic hydrocarbons. Current levels (μg/g wet wt) of aromatic hydrocarbons in fish (Douabul et al., 1984, 1987) from various coastal regions of the Arabian Gulf vary and range from about 1 to 4 in UAE (Khan et al., 1995; Shriadah, 2001), 1 to 70 μg/gm body weight in Kuwait (Al-Hassan et al., 2000), about 2.5 in Oman (Emara, 1990; Brady et al., 1993) and about 53 in Qatar (El-Deeb and El-Ebiary, 1988; Aboul-Dahab and Al-Madfa, 1993) (Table 1). In benthic fish, such as catfish *Arias bilineatus*, one-third of these hydrocarbons are PAH (Al-Hassan et al., 2001). The concentration in liver, gonads, and brain can be very high (up to 200 μg/g wet wt) (Table 2) (Al-Marri, 1994; Al-Hassan et al., 2001). Sharks from the Gulf can contain up to 12 μg/g and their newborns as much as 2 μg/g (Al-Hassan et al., 2000). Marine invertebrates from many phyla have also been shown to be contaminated with PAH (Lee, 1967, 1976, 1981; Anderson et al., 1974; Neff et al., 1976; Grahl-Nielsen et al., 1978; National Research Council, 1985; Malan, 1992; Lotufo, 1997; Gulec and Holdway, 1997, 2000; Al-Hassan et al., 2001).

*Corresponding author. *E-mail address:* maqkhan@uic.edu (M.A.Q. Khan).

Table 1. Concentrations of PAH in fish from the Arabian Gulf.

Fish species	Concentration (μg/g wet wt)	Location	Reference
Scolopsis binaculatus[1]	1.61	UAE, Dubai coast	Shriadah, 2001
Lethrinus minatus	1.56	UAE, Dubai coast	Shriadah, 2001
	2.71	UAE, Dubai Creek	Khan et al., 1993
Epinepehelus tauvina	2.50	Oman	Bradwi et al., 1993
Argyrops spp.	24.7	Qatar	El-Deeb and El-Ebiary, 1988
Malia spp.	53.4	Qatar	El-Deeb and El-Ebiary, 1988
Aria bilineatus	0.335	Kuwait	Al-Hassan, 2000
Saw-toothed Reef Shark	73	Kuwait	Al-Hassan, 2000
Gray Black Fin Shark		Kuwait	Al-Hassan, 2000
Males	11.9		
Females	7.93		
Unborn	2.2		
Milk Shark, newborn	20–60	Kuwait	Al-Hassan, 2000
Carpet Nurse Shark	7.93	Kuwait	Al-Hassan, 2000

[1] Concentration in tissues was: muscle = 0.70–3.57, gonads = 7.0–46.0, brain = 8.1–22.4. The concentration in *Arias thalassinus* was: muscle = 1.51, gonads = 77.0, and brain = 2.67 μg/g wet wt (Al-Marri, 1994).

The hydrocarbon residues in animals can exert various types of effects depending on their chemical nature, concentrations, biostability, exposure duration and stage of life cycle. These effects can be generalized as follows:

1. Acutely lethal: as observed in eggs, embryos, small and newborn, young, and spawning adults (Table 3) (Hyland and Schneider, 1976; Lee, 1977; von Westerhagen, 1987, 1988; McGurek and Brown, 1996; Carls et al., 1999).

Table 2. Petroleum hydrocarbon concentration in tissues of fish collected from Dubai Creek (Al-Marri, 1994).

Organ	Aryl hydrocarbon concentration (μg/g wet wt)[1]			
	Sparus sarba	*Arias thalassius*	*Pomocanthus maculosus*	*Scolopsis vosmeri*
Muscle	0.38–1.49	1.51	0.24–4.43	0.70–3.57
Gonad	2.54–55.64	77.12	3.82–38.47	7.00–46.00
Brain	5.36–86.66	2.67	0.57–226.5	8.10–21.44
Heart	2.71–54.10	3.92	2.70–9.27	6.06–32.25
Gill	0.40–10.24	11.21	0.06–1.04	0.58–2.76
Skin	0.85–5.61	11.39	0.72–15.64	0.75–1.55

[1] Habitat water concentration ranged from 1.20 to 8.59.

Table 3. Some general effects of petroleum hydrocarbons, as observed following the spills, on aquatic animals.

Concentration in water (mg/l)[1]	Observed Effects[2]
0.001	Bioaccumulation, tainting of edible tissues
0.010	Smell and taste affected, behavior (attraction to nest, food, etc., and repulsion from enemies, etc.) may be altered.
0.100	Metabolism (respiration, etc.) may be altered; growth and reproduction may be compromised; may cause malformations, cancer, etc.
1.00	Larval and juvenile mortality
> 10	Adult mortality

[1] These concentrations are common in estuarine and coastal waters.
[2] Refined oil is more toxic than crude. Effects increase in the following order: pelagic < subtidal < intertidal. Fish and decapod crustaceans, especially those from cold waters, are most sensitive.

2. Acutely or subchronically sublethal (toxic): such as pathological, behavioral (Lotufo, 1997), and physiological effects.
3. Prolonged and persistent: such as various life-taking diseases during life-cycle stages or continuing from generation to generation, such as cancers, neurological deficits, malformations, reproduction, etc.

Aquatic food-web depends on intricate interactions. An immediate effect of oil spill on photosynthetic organisms (Bott et al., 1976; O'Brien and Dixon, 1976; Bott and Crafford, 1985) can upset the population dynamics of other organisms, especially the zooplankton. Surface zooplankton takes up large amounts of PH from surface film. Aquatic animals, except mammals, are oviparous. Their eggs, larvae, embryos (Hose et al., 1981), and young and spawning adults (Singer et al., 1991; Hayes et al., 1992; Steering Committee, 1995; Gulec and Holdway, 1997; 2000) can be very sensitive to hydrocarbons (Table 3). For aquatic animals the sense of smell and taste are very important in responding to chemical messengers in water to maintain normal behavioral and physiological activities. PH, at parts per billion concentrations, can mimic these messengers and affect natural stimuli. These activities, which may be affected by low concentrations of volatile hydrocarbons include: nesting, mating, feeding and predation, escape from enemies, swimming, etc. The volatile hydrocarbons can also act as narcotic and necrotic agents to microfauna and may cause high mortality. The eggs and embryos and some small animals, such as bivalve mollusks (Stegeman and Teal, 1973; DiSalvo et al., 1975; Boehm and Quinn, 1976; 1977; Farrington et al., 1982; Widdows et al., 1982; Farley et al., 1991; Moore et al., 1991; Fisher and Foss, 1993), crustaceans (Alden and Butt, 1987; Malan, 1992; Lotufo, 1997; Woodhead et al., 1999; Gulec and Holdway, 2000) sipunculids (Rice et al., 1976; Lee et al., 1980), and polychaet worms (Lu et al., 2004) cannot withstand even parts per billion concentrations of hydrocarbons, which are not uncommon following spills (Table 1) or in highly trafficked waters. Following the exposure of embryonic or later stages to petroleum hydrocarbons, the surviving individuals may have their behavior altered and may not be able to reproduce normally (Von Westernhagen et al., 1987).

Some of these animals can accumulate hydrocarbons in their tissues, especially PAH and thus can be subjected to chronic effects. The studies in the Gulf have shown impact of hydrocarbons on Ostracod and barnacle crustaceans (Brady et al., 1993). The effects on scope for growth and lysosomal latency in mussels have also been reported from various other oil-polluted areas.

The effects of hydrocarbons depend on how much of these hydrocarbons are picked up, metabolized, retained, and excreted by the organism. Microorganisms are widely used for oil degradation, while fish species are commonly used to study their effects (Bobra et al., 1978). The fate and effects of PAH have been well studied in several marine and freshwater fish species. US Environmental Protection Agency (USEPA) has included 16 of the PAH on its priority list, whose toxic effects need to be investigated.

The surface organisms are most affected after the oil spill and intertidal PH can affect many spawning fish. The water-accommodated fraction, sediments, and their organic extracts are commonly tested for acute toxicity/sensitivity of animals. In addition to model experimental animals, some of these toxicity assays use commercially available "tox" systems (Papadopoulou and Scamara, 2002) and/or available cell lines, or biochemical markers.

Most of the PH end up in bottom sediments and organisms (Grahl-Nielsen et al., 1978); and about one-third of the sediment hydrocarbons are PAH. The chronic toxicity of sediments is site-specific because the nature and proportion and concentrations of various PAH may vary at different sites (Whipple et al., 1978). Also, the presence of other sediment contaminants is important in assessing the PAH caused chronic toxicity.

Chronic toxicity includes assays for carcinogenic, embryonic, developmental, and reproductive effects. These studies use whole animals, cell lines, and biochemical markers. Since 2,3,7,8-Tetrachlloro dibenzo(ρ)dioxin (TCDD) exerts its effects via the cytosolic Ah receptor (Hankinson, 1995), which is also the receptor of PAH metabolites, toxicity of sediments is usually expressed as dioxin-equivalent (Villeneuve et al., 2002). This chapter will focus only on the relationship of sediment PAH and liver carcinogenicity in fish. Most of the information will pertain to wild populations, while laboratory experimental data for this relationship will be included also. We can learn from the existing knowledge of the fate and effects of oil hydrocarbons gathered in other countries. This can help us in designing studies to investigate if similar effects can be possible in fish from the Gulf. Most of this work has been carried out in USA, which alone uses more petroleum than rest of the world put together, but where the combustion sources contribute 10-time more PAH than oil spills (Stroker and Seager, 1976). Urban and industrial activities can contaminate rivers, which may contribute to as much as 25% of total marine PH.

2. Cancer in North American fish species

2.1. PAH concentration in sediment and fish and occurrence of liver tumors

In most highly contaminated waterways and lakes the concentration of PAH can be as high as 100 μg/g dry weight of the sediments, while it is negligible in remote waterways/creeks/lakes (Tables 4 and 5). This is true for Great Lakes Region in Ohio, Michigan, and Illinois and other coastal areas of the United States of America. The concentration of PAH

Table 4. Concentrations of PAH in sediments of some selected US harbors.

PAH	Useless Bay	Concentration (mg/g dry wt) Washington		Massachusetts Boston H.
		Duwamish Waterway	Eagle H.	
2-ring	0.0027	0.340	0.260	15
3-ring	0.0049	1.800	0.330	3.500
4-ring	0.013	5.900	1.100	6.800
5-ring	NIL	0.620	0.510	1.600
Total PAH	0.021	8.200	2.200	26.000

Five-inch sediment from the top was generally used for analysis of PAH (data have been modified from various sources).

Table 5. Relationship of sediment PAH concentration with bile fluorescence in fish from Ohio.

Location	Sediment PAH[1] (mg/g dry wt)	BaP in bile[2] (μg/μl)	DNA–BPDE-adduct[3] (nmole/mole DNA)	Ref.
Ohio				
Black River	179.80	–		Dunn et al., 1987
Bullhead	–	10.10–104.70	29	
Carp	–	102.1–354.1		
Old Woman Creek	1.17	–		
Bullhead	–	0.70–71.70	NIL	
Buck Eye Lake	0.356	–		
Fish Hatchery	–	–		
Bullhead	–	0.00–4.90	NIL	
Washington				
English sole				Stein et al., 1993
Duwamish W.	8.20[1]	1300 + 800[2]	28.0 + 2.80	
Starry flounder			14 + 4	
Everett H.	2.200	450 + 240	17.0 + 7.60	
Useless bay	0.021	67 + 45	<0.20	
Boston				
Boston Harbor	26.00	–		
Winter flounder	–		9.0 + 7.8	
Ontario				
Brown bullhead				Hayes et al., 1990
Hamilton H.	38.52	54		
Oakville creek	0.10	16		
South Carolina				
Red drum				Reichert et al., 1998
Ashley R.	10.000	2000[3]	110	
Cooper R.	0.500	1000[3]	100	

1 Data from Fabacher et al. (1998) and Baumann et al., 1990.
2 Data from Varanasi et al., 1986, 1989.
3 Per mg protein.

Table 6. Concentration of pyrene and BaP in benthic invertebrates from Great Lakes (Fucik et al., 1995).

Lake	Concentration (μg/g dry wt)			
	Oligochaets		Amphipod[1]	
	Pyrene	BaP	Pyrene	BaP
Michigan	0.240–3.60	0.15–3.20	3.60–10.40	1.10–3.70
Erie	0.150–0.350	0.04–0.12	–	–
Ontario	–	–	–	0.33–0.90

[1] Amphipod = *Pontoporeia havoi.*

in benthic fish (Table 5) and invertebrates (Table 6) can be several times that in the sediments. The nonmigratory benthic animals are less active and have restricted home range. Their eggs can absorb and accumulate much higher concentrations of these hydrocarbons than the pelagic animals and their larvae.

More than 41 species of fish from various parts of USA and Canada have been reported to exhibit epizootics of hemic, neural, connective tissue, gonial, and liver cancers (Harshburger and Black, 1990). More than 14 of these species show hepatic and epidermal neoplasms, which are related to chemical pollutants in their habitat. The fish species with epizootic liver cancer are all benthic (Table 7) and are prevalent in population in waters highly contaminated with chemicals mainly PAH.

Some of the well-documented cases of PAH-induced hepatic tumorogenicity in fish reported in North America include locations, such as the Puget Sound–Duwamish Waterway (Seattle, Washington) (Sonstegard and Leatherland, 1984; McCain et al., 1987; Myers et al.,

Table 7. Hepatocytic and *Cholangiocytic neoplasia* in Fish (Harsburger and Black, 1990; Black and Baumann, 1990).

Habitat	% of fish with cancer	Ref.
Brown Bullhead (*Ictalurus nebulosus*)[1]		Rice et al., 1986; Black, 1983, 1988; Hayes et al., 1990
Black River, Ohio	15–35	
Cuyahoga River, Ohio	24	
Lake Erie and tributaries at Niagra	15	Black, 1983
Deep Creek Lake, Michigan	5	Dawe et al., 1964
Lakes in Pennsylvania	0–3	
Silver Stream reservoir, New York	5	
Muscong River, Michigan		
Fox River, Illinois	5	
White Sucker (*Catostomas commersoni*)		Black, 1988; Cairns and Fitzsimons, 1988a, 1989

(continued)

Table 7. (continued)

Habitat	% of fish with cancer	Ref.
White Sucker (*Catostomas commersoni*)		Black, 1988; Cairns and Fitzsimmons, 1988b
Lake Ontario, Canada	12	
Eastern Lake Erie and Torch lake, Michigan	12	
Niagra River, New York; Lake Michigan		
Surger (*Stizostedion canadensa*) and Walleye (*S. vitreum*)		Black and Baumann (1990)
Torch Lake and Portage Lake, Michigan	25	
English sole (*Pleuronectes vetulus*)		Harshburger and Black, 1990; Varanasi et al., 1986
Puget Sound, Washington	10–35	
Vancouver harbor, Canada	0	
Atlantic tomcod (*Macrogadus tomcad*)		Smith et al., 1979; Cormier et al., 1989
New York		
Winter flounder (*Pseudopleuronectes americanus*)	High	Murchelano and Wolke, 1991; Gardner et al., 1989
Boston Harbour		
Red Croaker		
Charleston Harbour	High	Overstreet, 1988
Black bullhead		Grizzle, 1985
Tuskegee, Alabama		
White croaker (*Genyonemus lineatus*)		Malins et al., 1987a; Myers et al., 1991
Northern pike		
Great Lakes		Black, 1988
White perch		
Baltimore Harbour, Chester River, Chesp. Bay		Harshburger and Black, 1990; May et al., 1987
Lake trout		
New York		Hendricks et al., 1984

1990), Black River and Kuyahoga River (Ohio) (Baumann and Mac, 1988), Boston Harbor (Massachussetts) (McMahon et al., 1988a, 1990), Charleston Harbor (South Carolina) (Overstreet, 1988), and Hamilton Harbor (Ontario, Canada) (Metcalfe et al., 1989). The fish from PAH-polluted waters show high incidence of liver tumors as compared with the populations from uncontaminated waters in the same region (Landahl et al., 1990) (Table 7).

Great Lakes Region: Benthic fish, brown bullhead *Ictalurus nebulosus* from the highly contaminated Black River and Kuyahoga River in Ohio (Hinton et al., 1981; Baumann

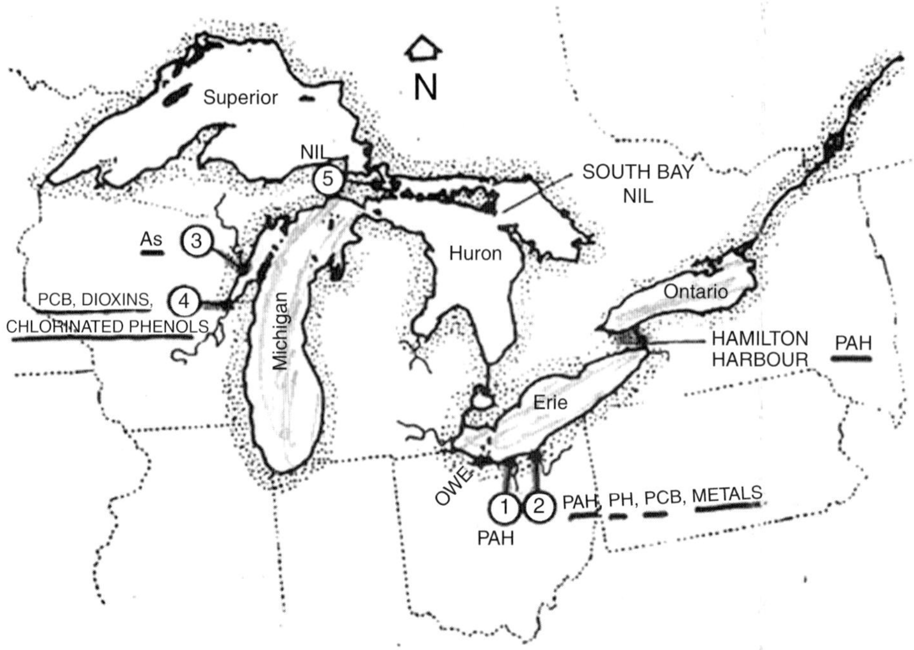

Figure 1. Map of the Laurentian Great Lakes of North America, indicating sites from which sediments were collected. 1. Black R.; 2. Cuyahoga R.; 3. Menominee R.; 4. Fox R.; 5. Manuscong L. The words underlined indicate chemical contaminants prevalent in the area (PAH, PH, PCB, metals, arsenic, dioxins, chlorinated phenols).

and Mac, 1988; Vogelbein et al., 1990) and from Fox River, Illinois (Baumann et al., 1982, 1987) (Figure 1), and Hamilton Harbor, Ontario (Metcalfe et al., 1989; Hayes et al., 1990) show a high prevalence of liver cancer. Up to 35% of the population of brown bullheads in Black R. show liver cancer (Table 7) (Baumann et al., 1984, 1990). While other creeks

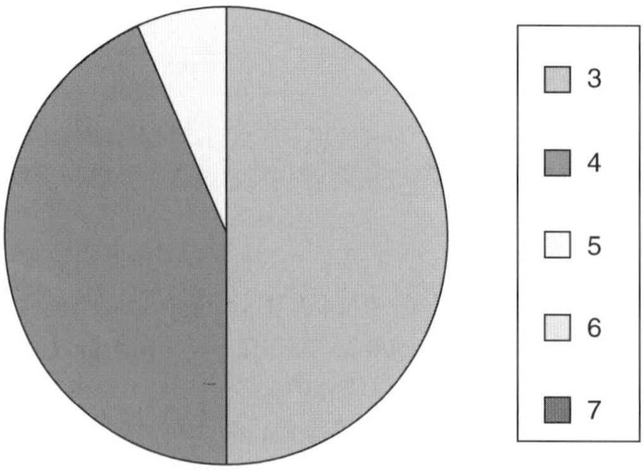

Figure 2. Proportion by age group of bullheads greater than 250 mm long collected in Black River. Numbers in legend shows the age of fish in years (Baumann et al., 1990).

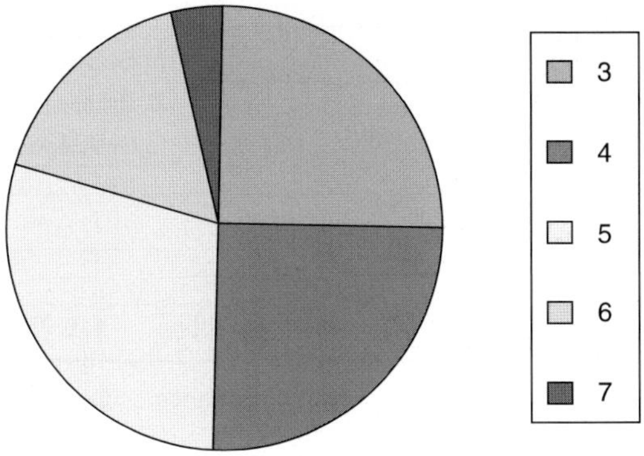

Figure 3. Proportion by age group of bullheads greater than 250 mm long collected in Old Woman Creek (Baumann et al., 1987).

show no cancerous fish and are inhabited with more than 40% of individuals at and above 5 years of age (Figure 2) that in Black R. shows only about 3% of its population to contain 5-year old individuals (Figure 3) (Baumann et al., 1987). The cancer takes about 4 to 5 years to kill the fish. Similar prevalence of cancerous fish has been reported for Kuyahoga River. Hamilton Harbor is also highly contaminated with PAH (and other carcinogens) and benthic fish show prevalence of liver cancer (Table 7) (Metcalfe et al., 1988). Rivers and lakes by urban and industrial sources are other locations where liver cancer in fish is very prevalent (Rice et al., 1986; Black, 1983, 1988).

Northwestern USA: The benthic fish English sole *Parophrys vetulus* from Puget Sound–Duwamish Waterway show a high prevalence of liver neoplasia as compared with those in less contaminated waters (Myers et al., 1991).

Northeastern USA: Winter flounder *Pseudo-pleuronectes americanus,* from highly contaminated Boston Harbour, show high incidence of liver tumors (Table 7) (Murchelano and Wolke, 1985, 1991; Gardner and Pruell, 1988; Gardner et al., 1989).

Southeastern USA: Red and black croaker from Charleston Harbor, South Carolina and from rivers in Alabama, which are highly contaminated with PAH, show high prevalence of liver tumorogenicity (Overstreet, 1988; Reichert et al., 1998). Black croaker populations are highly susceptible to high levels of PAH in sediments (Figure 4).

3. Fish cancer as an indicator of ecological and health effects

The prevalence of epizootic of liver cancer is directly related with environmental chemicals, mainly the PAH. The basin of Lake Erie and Ontario, and most critically that of Lake Michigan, where fish liver cancer has been demonstrated, are considered to be highly polluted with environmental chemicals, especially PAH. This is associated with observable ecological and health effects in this region, typically in the Lake Michigan basin (USEPA, 1995). Fish-eating birds and mammals and humans are badly affected by these chemical contaminants. Cancer, malformations, lower birth weights, reproductive failure (sex reversal in mammals, birds, and fish) (Hileman, 1988; Colborn et al., 1997),

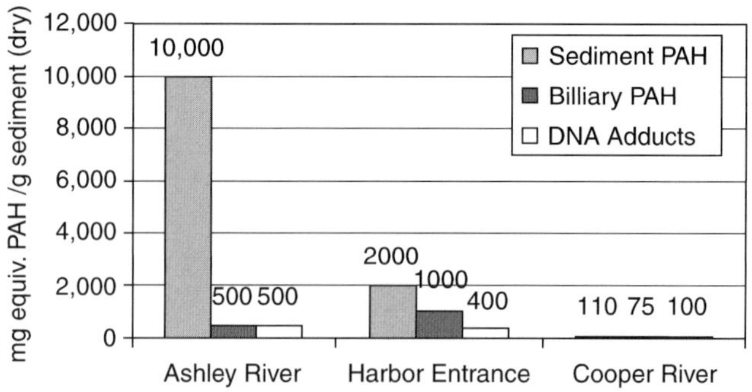

Figure 4. Sediment and billiary PAH and DNA adducts in Red drum from Charleston Harbor, South Carolina and uncontaminated Cooper River (Reichert et al., 1990).

etc. are more prevalent in this region than any other part of the world (Table 8) (USEPA, 1995). Thus, epizootic of fish liver cancer can serve as an indicator of contamination with PAH and possibly other carcinogens and of risks to health and ecosystem (Figure 5). Women in the reproductive age are advised not to consume large fish from Lake Michigan and not to breast feed their infants (Table 8). This is one of the most important example of

Table 8. Some ecological and health effects of pollutants in Lake Michigan basin (Chemical and Engineering News, 1988).

Humans	The risk of contamination is high if they eat fish from Lake Michigan. The incidence of miscarriages and birth defects is higher in this region while the birth size and gestational age are reduced. Neonates show neurobehavioral deficit by being hyperactive.
	There is an advisory for lake trout: nursing mothers, pregnant women, women anticipating conception, and children under 15 years of age "do not eat trout from Lake Michigan more than 32-inch long." MEN AND WOMEN: "DO NOT EAT LARGE TROUT."
	Overall cancer death rate, heart disease, birth defects are higher in this region than many parts of USA. US congress has declared this basin as a national model of contaminants affecting human health!
	PCB, chlorinated insecticides, PAH, and metals present in water and sediments. The data on the levels of various PAH in air, water, bottom sediments, and invertebrates and fish are available.
Fish	Trout and salmon show high incidence of liver tumors, goiter, and impaired metabolism and reproduction (secondary sex characters altered). In Ohio and Ontario fish in urban waterways and lakes have higher incidence of liver tumors. The life expectancy of these fish is reduced affecting their size distribution.
Birds	Double-crested cormorants, Caspian tern, Foster's tern, and herring gulls have defective embryos. Cormorant: in Green Bay: hatchlings failure and those with cross-bills are more common, sexual reversal (feminization of males) is supposed to be related with hormonal effects (also seen in fish).

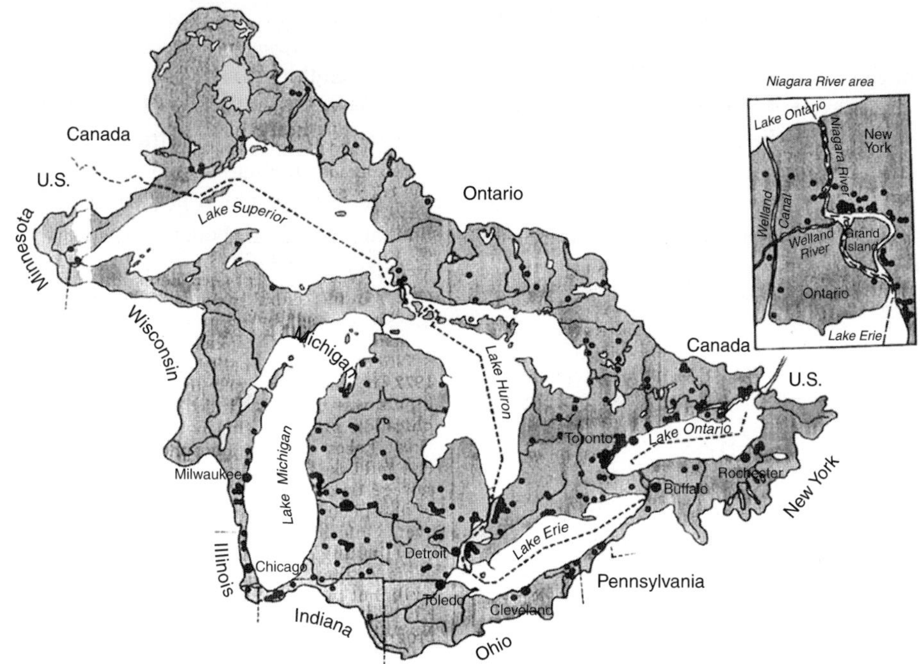

Figure 5. Great Lakes basin showing sites (*) where toxic chemicals are causing health and ecosystem effects (US EPA, 1995).

the effects of aquatic pollutants, especially PAH, PCBs, etc., on animals at the top of the foodchain (fish, frogs, birds, mammals, humans), via their trophic transfer (Black and Baumann, 1990; Colborn et al., 1997).

Several other polluted rivers and lakes in Great Lakes Region exhibit similar increases in cancerous populations of bottom fish. This extends all the way to the polluted Canadian lakes and rivers (Black, 1988; Metcalfe et al., 1988). The epidemiology of these diseases signifies the PAH and other toxics etiology.

The presence of PAH in commercial seafood in USA (Table 9) (Stegman and Teal, 1973; Stegeman, 1989) indicates that intake of other food and water and other routes of exposure may have to be considered in any extrapolation of fish data.

Table 9. Levels of benzo(a)pyrene (BaP) in commercial seafood in USA (Stegeman, 1989, 1991).

Seafood Source	Number of samples examined	BaP concentration (mg/gm wet wt)
Mollusks	34	5.90 ± 8.30
Crustaceans	19	2.30 ± 2.30
Fish (cod)	14	1.30 ± 0.09

3.1. PAH and liver carcinogenicity in fish

The field studies clearly indicate that some (but not all) benthic fish in PAH- contaminated waters have very high prevalence of liver neoplasia (Metcalfe et al., 1984). That PAH in sediments can be responsible for liver carcinogenicity has been demonstrated by testing the sediment extracts (PAH) for tumorogenicity using mammalian models. For example, sediment extracts from Black River, Ohio when painted on the skin of nude mice showed the rate of tumorogenicity that compared with that caused by benzo(a)pyrene (BaP) (Figure 6) (Black et al., 1985). Sediment extracts injected into the fish or mice have also produced liver tumors (Fabacher et al., 1988). The short-term *in vitro* assays, which are

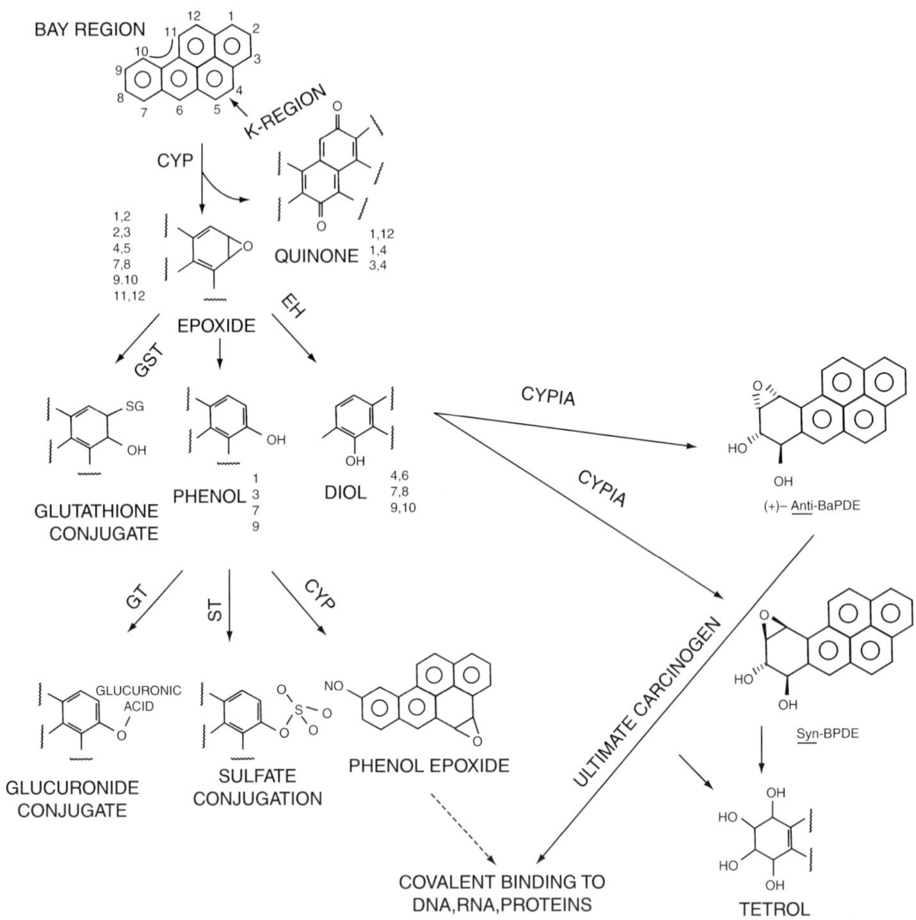

Figure 6. Major pathways of metabolic activation and detoxification of benzo(a)pyrene (BaP) by microsomal cytochrome P-450 (CYP), epoxide hydrolase (EH), and UDP-glucoronic acid transferase (GT), and by cytosolic glutathione-transferases (GST) and sulfotransferase (ST). BaP is attacked in Bay-region by CYPIA and in K-region by CYPIIB. *Anti*-BaP-7,8-diol-9,10-epoxide (BaPDE) binds covalently to, and damages, DNA (initiation of tumorigenesis).

Table 10. In vitro mutagenicity of the extracts from Hamilton Harbor and Oakville Creek in Ontario, Canada (Metcalfe et al., 1990).

Test	Source of sediment	
	Hamilton Harbor	Oakville Creek
Ames Test	Positive	Negative
Sockfry microinjection[1]	Positive	Negative
Mouse C3H1OT1/2	Mouse C3H1OT1/2	Mouse C3H1OT1/2

[1] Rainbow trout *O. mykis.*

commonly used for mutagenicity testing of pollutants, show strong positive response for the sediment extracts (Table 10–12) (Metcalfe et al., 1989).

In addition to measuring PAH in fish and bile (Varanasi et al., 1986) other indicators (assays that eventually lead to initiation of tumorogenecity) of fish exposure to PAH are: induction of hepatic cytochrome P4501A (AHH) activity and concentration (Payne, 1976; Payne and May, 1979; Bend et al., 1979; James et al., 1976, 1979, 1988; Collier et al.,

Table 11. Some of the known genotoxic sites of carcinogenic PAH in fish (Reichert et al., 1998).

Fish	Target gene	Reference
Tomcod	Activation of the k-*ras* oncogene CYPIA mRNA	Wirgin et al., 1989
English sole	ki-*ras* b cDNA cloned and sequenced Areas of amino acid sequence conservation include Codons 12, 13, and 61 positions	Peck-Miller et al., 1998
Rainbow trout	Mutational activation of an expressed k-*ras* gene by PAH	Fong et al., 1993

Table 12. Ames mutagenicity test (McCann et al., 1975) (using TA 100 and TA 98 *Salmonella* strains) with extracts of sediments from Hamilton Harbor (PAH = 38.52 μg/g) and uncontaminated Oakville Creek and South Bay.

Location	Treatment			
	TA 100		TA 98	
	With S-9	w/o S-9	With S-9	w/o S-9
Hamilton Harbor				
2.5 g equivalent	202	203	61	64
Spontaneous	102	105	33	39
Oakville Creek				
2.5 g equivalent	92	15	41	37
South Bay	93	57	40	38

Data from Metcalfe et al. (1990).

Table 13. Various types of effects of PAH on fish used as biomarkers for their exposure.

1	Induction of cytochrome P-4501A1 (Stegeman and Leech, 1991, Gocksoyr et al., 1991).
2	^{32}P-labelling of PAH-DNA adduct formation. (Stein et al., 1993).
3	CYPIA mRNA (Stegeman and Leech, 1991).
4	Lysosomal morphology indicator of histopathological changes. Lysosomal membrane stability, when compromised, can cause degeneration leading to adenomas and carcinomas (Kohler et al., 2002).
5	Peroxisome proliferation. This is massive and reversible. Peroxisome volume density is increased and acetyl co-enzyme A oxidase activity is monitored (Amemerschlaeger et al., 2004).
6	Vitellogenin and zona radiata proteins, which are synthesized in liver and affect maturation of eggs (Trescot et al., 1988).

1995), formation of carcinogenic BaP metabolites (Osborne and Crosby, 1987; Guengerich, 1992) and their adducts with DNA, assay of mRNA due to transcriptional activation of CYP4501A1 gene by PAH (Table 13) (Stegeman, 1989; Metcalfe et al., 1990; Stegeman and Lech, 1991). However, other chronically PAH-exposed fish species may be refractory to *CYPIA1* gene induction (Eiskus et al., 1999) even tough they may produce as much BPDE–DNA adducts (Stein et al., 1987, 1993).

The benthic fish from PAH-contaminated waters show high levels of PAH and their metabolites in their tissue, especially in bile (~700 and 900 ng/ml bile, which is 2 to 27 times higher than in the control) (Table 14) (Stein et al., 1987; Varanasi et al., 1983; 1986; 1989a; 1989). However, various benthic fish in PAH-contaminated waters are not susceptible to liver tumorogenicity. For example, liver tumors have been demonstrated in English sole, winter flounder, black croaker, brown bullheads but not in other benthic fish from each of these respective locations. These differences in susceptibility to cancer have been attributed to differences in the pathways of biotransformation of PAH (Guengerich, 1988, 1992) between resistant and susceptible fish (Stein et al., 1993; Wirgin et al., 1991). It has been shown, using BaP as a model, that susceptible fish produce more of the BaP metabolic products, which form stable DNA adducts. The later cause irreparable DNA damage and this initiates mutagenicity in affected hepatocytes leading to neoplastic growth. This has been demonstrated in laboratory studies using the

Table 14. A comparison of fate of an intraperitoneal dose of ^{3}H-BaP (7.9 μmole/kg in acetone) in English sole and Sprague–Dawley male rat (Nishimoto and Varanasi, 1985, 1986).

Animal[1]	BaP concentration		CBI[2]
	(Liver pmole equiv./mg)	Bile (pmole equiv./mg)	
English sole	200 ± 40	4600 ± 600	2.10 ± 0.50
Rat	6.0 ± 5.0	–	0.30 ± 0.10

[1] *In vitro* AHH activity = 190 ± 60 fish and 600 ± 42 rat and formation of 7,8-dihydro-BaP-diol = 50 ± 10 in fish and 28 ± 1 pmole/min/mg protein in rat. Data from Varanasi et al. (1986).

[2] Covalent binding index = umole BaP bound per mole DNA/umole administered per kg body wt.

resistant and susceptible fish models, such as English sole and starry flounder (Myers and Rhodes, 1988). Other experimental fish species, which have been shown to develop hepatic neoplasia by model PAH carcinogens, include Coho salmon (Ostrander et al., 1988, 1989), steelhead trout (Kocan and Landolt, 1984), Rainbow trout (Hendricks et al., 1984, 1985; Metcalfe et al., 1988; Fong et al., 1993), Japanese medaka (Hawkins et al., 1990), Guppy (Hawkins et al., 1989), etc.

The model rodent carcinogen, BaP, causes skin and liver tumors in experimental rodent and fish species, respectively. This PAH is oxidized at various sites (Figure 7). In K-region the oxygenation by P4502B (see Khan, 1980) and further metabolism leads to noncarcinogenic products. On the contrary, in the Bay region the oxygenation of BaP leads to the formation of genotoxic metabolites, which are the ultimate carcinogens. The attack in Bay region by the liver microsomal P4501A produces 7,8-epoxide of BaP (Table 14). This epoxide is attacked by the microsomal epoxide hydrolase forming 7,8-dihydrodiol of BaP This diol is again attacked by P450 to form 9-10-epoxide. This benzo(a)pyrene-7,8-diol-9,10-epoxide (DPDE) has *syn* and *anti* isomers. The *anti*-BPDE is the ultimate metabolite that binds covalently with the d-guanine of DNA (Table 14). The covalent binding index is higher in sole than rat. The BPDE–DNA adduct, if not broken down to DNA (DNA repair) (Table 14), initiates mutation, altered gene expression, and proliferative cell growth. At least this is the first step in the DNA damage and mutagenicity. BaP inside the hepatocyte is translocated to the nucleus where it causes transcriptional activation of *CYPIA1* gene whose mRNA then increases the synthesis of P4501A protein. PAH, thus, induce the synthesis of the P450 isozyme that activates them to carcinogenic metabolites. *In vivo* and *in vitro* activities and contents of P451A from PAH-exposed fish are higher than in unexposed fish. In addition to P4501A other enzymes, such as UDP-glucuronyl transferases (UGT), may be important in cancer susceptibility of fish. For example, in human lymphoma polymorphism the human phenotypes with low levels of UGT are more susceptible to DNA damage (Hu and Wells, 2004). Since P4502B cannot be induced in fish and aquatic amphibians (Khan et al., 1998), this makes them extremely sensitive to BaP and other PAH carcinogens.

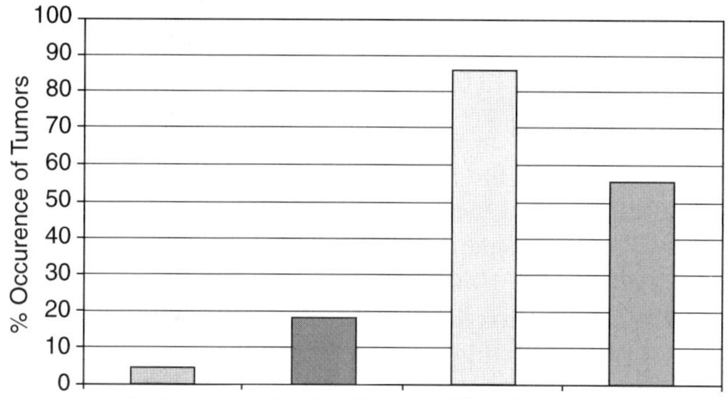

Figure 7. Effect of sediment extracts from contaminated and uncontaminated sites in Ohio on skin tumors in Swiss mice (30 weeks) as compared with BaP (Black et al., 1985).

4. English sole as a fish model for BaP carcinogenicity

4.1. Mechanisms of action of PAH carcinogens: BaP metabolism

PAH are procarcinogens and need to be activated by liver enzymes, as described above with BaP, to genotoxic metabolites, which damage DNA and initiate mutation and carcinogenicity. BaP metabolite in heapatocyte binds specifically to the cytosolic arylhydrocarbon receptor (AhR) (Billiard et al., 2002) causing conformity changes that affect interaction with Arnt cofactors. The AhR translocates the BaP metabolite inside the nucleus where its complex with Arnt regulates transcription of CYPIA1/2 gene (which has low level constitutive expression and causes robust increase in response to AhR ligand) through specific (dioxin-responsive) elements. Regulation of CYPIA1 relies on a multitude of factors. Both endproducts of induction, mRNA and protein (P4501A1/2) are commonly used as indicators of exposure to PAH. The coordinate binding of the BaP metabolites to DNA (BPDE–DNA adduct) may persist, resist repair elements and lead to DNA damage and mutation of the affected cell. Not all hepatocytes show similar dose–response relationship. It is postulated that a hybrid switch response may be at work with a binary switching mechanism responsible for whether the gene is active or not and a second rheostat type switch governing variability in transcriptional output (Cohen et al., 2004; Huff, 2003; Tomatis, 2002).

DNA adducts. Radiolabelled BaP is either injected into the body cavity or muscle of the fish or fed orally. This is followed by time–course study of the metabolism of BaP. Most of the biotransformation takes place in liver, initially by microsomal P450 oxygenases and then epoxide hydrolase. Liver enzymes attack the hydroxylated metabolites further. It conjugates by attaching glucuronic acid or glutathione or sulfate to the hydroxylated metabolite via liver enzymes. The resulting conjugates of the hydroxylated metabolites are water soluble and can be excreted out of the liver via bile. The bile can also contain free metabolites of BaP as well as unmetabolized BaP. These metabolites and BaP can be quantified by measuring the fluorescence in bile. The distribution of BaP and metabolites is measured by using various analytical techniques (Maccubbin et al., 1988; Krahn et al., 1984, 1987; Stein et al., 1993). These studies can help in understanding the fate of BaP and possibly other PAH. Since, from the point of view of carcinogenicity, we are interested in the formation of BDPDE and its DNA adducts (Malins and Haimanot, 1990), the rate of these reactions is used to estimate the potential of carcinogenicity of BaP.

The half-life of injected BaP is shorter in English sole (much shorter in fish from PAH-contaminated site) than in the flounder. Twenty-four hours after injection, 1% of the dose was present in liver and metabolites included phenols, quinones, dihydrodiols, and their conjugates. The bile contained 20% of its radioactivity in form of diols while liver and gonads contained only 10%. The major unconjugated metabolites were BP-9, 10-dihydroxydiol, BP-7,9-dihydroxydiol, and 1- and 3-hydroxy BP. The concentration of conjugated diols in bile was higher in English sole than in starry flounder from contaminated sites (Varanasi et al., 1989). In English sole BPDE–DNA adduct formation *in vivo* reached a maximum within 2 days after the administration of BaP (100 umole/Kg BW, im) with a half-life of 11 days (Table 14). About 36% of the adduct still remained intact up to 84 days (Nishimoto, 1986; Stein et al., 1993). In the juvenile sole adduct can persist for at least 2 months (Varanasi et al., 1986). Similar adducts of aflatoxin B-1 in Coho salmon (Bailey et al., 1988) and of dibenzocarbazole in English sole (Stein et al., 1993) were quite persistent. In Brown bullhead the BPDE-adduct formation is slow but adducts persists as long (Lotlikar et al., 1984; Nishimoto and Varanasi, 1986; Baumann et al., 1990). Compared to fish the half-life of BPDE–DNA adduct is only 8 h in rats (Varanasi et al., 1986). BPDE–DNA adducts formation, which if not repaired, can cause mutation of the K-*ras* protooncogene and, thus, the initiation of tumorogenicity. The differences in the sensitivity of fish to cancer are mostly related with these metabolic events.

Ras oncogenes, with activating point mutations in codons 12, 13, and 61, are found in greater frequency than any other gene in human tumor cells (Anderson et al., 1992; Barbacid, 1999), ki-*ras* genes are present in chemically-induced hepatic tumors in Atlantic tomcod (Wirgin et al., 1989), winter flounder (McMahon et al., 1988, 1990), English sole (Peck-Miller et al., 1988), rainbow trout (Chang et al., 1991; Hendricks et al., 1994; Omer et al., 1995), and medaka (Torten et al., 1996). Other mechanisms of tumorogenicty and apoptosis may include induction of mitotic arrest and p53 activation (Shin et al., 2004).

Peroxisome proliferation. Fish (and bivalve mollusks) seem to be sensitive to environmental chemicals, which cause peroxisome proliferation (Cancio and Cajaravillem, 2000; Ammerschlaeger et al., 2004). However, the tumorogenesis via peroxisome proliferation is not known in fish and invertebrates (Cajarville et al., 2003).

Free radicals. Various reactive radical intermediates in BaP metabolism can cause DNA damage (Cavlieri and Rogan, 1992; Maccubbin et al., 2000; Laramaira and Livingston, 1997).

Lysosomal membrane damage. This biomarker is used by OECD as one of the bioindicators of chemical damage (Kohler et al., 2002).

4.2. Other effects of PAH on fish

The acute toxicity of PH is generally attributed to their narcotic effects (French-McCay, 2003). This is considered to be due to their partitioning into cell membranes especially in the nervous tissue causing depression of nervous system (Barron et al., 2003). PAH are unlikely to act as narcotic agents in early life-stages of fish that have been chronically exposed as embryos. Sublethal effects caused by embryonic PAH exposure include edema of the yolk sac and pericardium, hemorrhaging, disruption of cardiac function, binding to AhR and CYPIA induction, mutations and heritable changes in progeny, craniofacial and spinal deformities, neuronal cell death, anemia, reduced growth, and impaired swimming (Marty et al., 1997; White et al., 1999; Billiard et al., 2002; Barron et al., 2003). Chrionic toxicity of PAH in Pacific herring and pink salmon show that PAH with high relative AhR affinity did not contribute substantially to the observed early life-stage toxicity. Narcosis contributed to embryo mortality and to be predominantly controlled by the concentration of naphthalenes (Barron et al., 2004).

Aromatic hydrocarbons and their metabolites and DNA adducts, accumulating in gonads, can enter sperms and eggs of English sole adults (Varanasi et al., 1986) and juveniles. The eggs and alevins of steelhead trout retain BaP (Kocan et al., 1985, 1987; Westernhagen et al., 1987). Since the eggs and embryos of fish are pelagic, the parental exposure of developing embryos plus the exposure of fry through surface water can be fatal. The embryos of these fish are sensitive to very low concentrations of water-soluble fractions of crude oil (Table 3). PAH can reduce recruitment and alter gonad size, gametogenesis, levels of sex steroids (Khan and Kiceniuk, 1984; Trescot et al., 1988; von Westernhagen, 1988; Johnson et al., 1995), and increased malformations in larval Pacific herring (*Clupea pallasi*) (McGurk and Brown, 1996). The coho salmon embryos exposed to BaP show subtle effects after hatching and the behavior of the fish can be compromised including upstream orientation, swimming, foraging, and mating (Ostrander et al., 1988, 1989, 1990). BaP affected alevin and fry behavior, caused temporal differences in hatching and the ability to perform characteristic behavior associated with emergence. Embryonic exposure to water-soluble fraction of PH ($\sim$1 ppb) and oil-contaminated gravel caused yolk-sac edema, hemorrhages, related growth and occasional spinal deformity (May et al., 1987; Marty et al., 1997; Carls et al., 1999), PAH were most toxic (Heintz et al., 1999).

Exposure of eggs of mummichog (*Funfulus heteroclitus*) to oil-contaminated sediments resulted in retarded growth and development, pericardial edema, haemostasis, hemorrhages, and spinal deformities (Couillard, 2002). Similar effects have been reported in other species, which spawn in intertidal zones, where most of the oil spill is stranded (National Research Council, 1985). These species include herring (*Clupea* spp.), smelt (*Osmerus mordax*), and capelin (*Mallotus villosus*) and mummichog (Carls et al., 1999; Barron et al., 2004). Concentrations as low as 4.5 ug/g sand can reduce body length in mummichog (Colliard, 2002).

Steelhead trout can store BaP in their nervous system and eye muscles (Kocan and Landolt, 1984; Kocan et al., 1985, 1987). High concentrations of PAH were present in brains of fish from UAE (Al–Marri, 1994).

Fish enzymes (P4501A) induced by PAH (James et al., 1976; 1979; Khan et al., 1998; Lee et al., 2001) can affect the metabolism and disposition of other endogenous steroids and foreign chemicals. The resulting alteration of hormones can affect homeostasis, spawning, and reproduction. Activities of P450E are negatively correlated with fertilization success in spawning starry flounder in San Francisco Bay area and white croaker *Genoyomus lineatus* in Los Angeles area (Malins et al., 1987a; Malins and Haimanot, 1990).

4.3. Seafood as a bioindicator of PH contamination

The presence of petrogenic PAH in water, sediments, air, and living organisms indicates the potential of human exposure. The exposure can be quantified by knowing the concentration and thus amounts to which Gulf residents are being and have been exposed.

The fish species sensitive to PAH carcinogenicity can be used as a bioindicator of PAH contamination (Black, 1984; Myers et al., 1991; Cooper et al., 1991). This may indicate the potential of harmful effects on ecosystem and humans (Cohen et al., 2004). However, the concentration of BaP, BPDE, and BPDE–DNA adducts (Dunn, 1991) in edible parts of fish and seafood do not indicate that consumption of fish and seafood from petroleum-contaminated sites poses significant threat, unless the fish come from highly polluted waters and are regularly consumed. In USA (Table 9) and possibly other countries, seafood and other food items and fluids may also contain carcinogenic PAH.

The total exposure of humans to PAH and other carcinogens should also take into consideration other routes of environmental exposures. Information about similar aspects of human cancer risk has been considered for the North American situation (Cooper et al., 1991; Dunn, 1991). In the case of fish the BPDE–DNA adducts are formed in higher amounts in cancer-susceptible fish than rodents (which are resistant to BaP hepatic tumorogenicity), which persist for several months. Extrapolation to rodents and then humans of fish bioindicator needs further elaboration. The experimental rodent data on chemically induced tumorogenesis may or may not invariably predict human cancer hazard (Tomatis, 2002; Huff, 2003; Cohen et al., 2004). Alternatively, each experimental tumor site should be treated individually to identify key events for any proposed mechanism of action as well as identification of data gaps. The limits of available data and negative findings need to be taken into consideration. For rodents the precise mechanism of action and dose–response for tumorogenicity should be precisely known. Whether the same mechanism and dose–response relationship exists in human makes cancer risk hazard extrapolations more realistic (IARC, 1983; Tomatis, 2002; Huff, 2003; Health Risk Assessment of Chemical Contaminants in Seafood, 1988; Steering Committee, 1991, 1995; Wilson, 2004; Cohen et al., 2004). It is not unwise to overestimate the potential risk of cancer and other diseases from all these exposures.

References

Aboul-Dahab, O., Al-Madfa, H., 1993. Oil pollution in Qatar coastal sediments. Environ. Poll. 81, 113–116.

Alden, R.W., Butt, A.J., 1987. Statistical classification of the toxicity of polynuclear aromastic hydrocarbon contamination of sediments from a highly industrialized seaport. Environ. Toxicol. Chem. 6, 673–684.

Al-Hassan, J.M., Afzal, M., Rao, C.V.N., Fayad, S., 2000. Petroleum hydrocarbon pollution in sharks in the Arabian Gulf. Bull. Environ. Contamin. Toxciol. 65, 391–398.

Al-Hassan, J.M., Afzal, M., Chava, V.N.R., Fayad, S., 2001. Hydrocarbon pollution in the Arabian gulf fish Arius bilineatus. Bull. Environ. Contamin. Toxicol. 66, 646–652.

Al-Marri, S., 1994. Petrogenic aromatic hydrocarbons in fish from Dubai. Biology Department, UAE University, Al Ain, UAE.

Ammerschlaeger, M., Beigel, J., Klein, K-U., Mueller, S.O., 2004. Characterization of the species-specificity of peroxisome proliferators in rat and human hepatocytes. Toxicol. Sci. 78, 229–240.

Anderson, M.W., Reynolds, S.H., You, M., Maronport, R.M., 1992. Role of proto-oncogenes activation in carcinogenesis. Environ. Health Perspect. 98, 13–24.

Anderson, J.W., Neff, J.M., Petrocelli, S.R., 1974. Sublethal effects of oil, heavy metals, and PCBs on marine organisms. In: Khan, M.A.Q., Bederka, J.P. Jr. (Eds), Survival in Toxic Environments. Academic Press, New York, pp. 83–124.

Bailey, G.S., Williams, D.E., Wilcox, J.S., Loveland, P.M., Coulombe, R.A., Hendricks, J.D., 1988. Aflatoxin B carcinogenesis and its relation to DNA adduct formation and adducts persistence in sensitive and resistant salmonid fish. Carcinogenesis 9, 1919–1926.

Barbacid, M., 1999. Ras gens. Annu. Rev. Biochem. 56, 779–787.

Barron, M.G., Carls, M.G., Short, J.W., Rice, S.D., 2003. Photoenhance toxicity of aqueous phase and chemically dispersed, weathered Alaska North Slope crude oil to Pacific herring eggs and larvae. Environ. Toxicol. Chem. 22, 650–660.

Barron, M.G., Carls, M.G., Heintz, R., Rice, S.D., 2004. Evaluation of fish early life-stage toxicity of chronic embryonic exposures to complex polycyclic aromatic hydrocarbon mixtures. Toxicol. Sci. 78, 60–67.

Baumann, P.C., Mac, M., 1988. Polynuclear aromatic hydrocarbons and tumors in brown bullheads from the Black and Cuyahoga Rivers - cause and effect? In: Niimi, A.J., Solomon, K.R. (Eds), Proceedings of the Fourteenth Annual Aquatic Toxicity Workshop, November 2–4, 1987, Toronto, Canadian Technical Report of Fisheries and Aquatic Sciences, No. 1607, pp. 142–143.

Baumann, P.C., Smith, W.O., Ribick, M., 1982. Hepatic tumor rates and polynuclear aromatic hydrocarbon levels in two populations of brown bullheads *Ictalurus nebulosus*. In: Cooke, M., Dennis, A.J., Fischer, G.L. (Eds), Polynuclear Aromatic Hydrocarbons: Physical and Biological Chemistry. Battelle Press, Columbus, OH, pp. 93–102.

Baumann, P.C., Smith, W.D., Parland, W.K., 1987. Tumor frequency and contaminant concentrations in brown bullhead from an industrialized river and recreational lake. Trans. Amer. Fish. Soc. 116, 78–86.

Baumann, P.C., Harshbarger, J.C., Hartman, K.J., 1990. Relationship between liver tumors and age in brown bullhead populations from two Lake Erie tributaries, The Science of the Total Environment 94. Elsevier, Amsterdam, pp. 71–87.

Bend, J.R., Ball, L.M., Elmamlouk, T.H., James, M.O., Philpot, R.M., 1979. Microsomal mixed-function oxidation in untreated and polycyclic aromatic hydrocarbon-treated marine fish. In: Khan, M.A.Q., Lech, J.J., Menn, J.J. (Eds), Pesticide & Xenobiotic Metabolism in Aquatic Organisms. Plenum, New York, p. 297.

Billiard, S.M., Hahn, M.E., Franks, D.G., Peterson, R.E., Bols, N.C., Hodson, P., 2002. Binding of polycyclic aromatic hydrocarbons (PAHs) to teleost aryl hydrocarbon receptor. Compar. Biochem. Physiol. 133B, 55–68.

Black, J.J., 1983. Field and laboratory studies of environmental carcinogenesis in Niagara River fish. J. Great Lakes Res. 9, 326–334.

Black, J.J., 1984. Aquatic animal neoplasia as an indicator for carcinogenic hazards to man. In: Saxena, J. (Ed.), Hazard Assessment of Chemicals — Current Developments. Academic Press, New York.

Black, J.J., 1988. Chronic effects of toxic contaminants in large lakes. In: Schmidthe, N. (Ed.), Toxic Contaminants in Large Lakes, Vol. 1. Lewis Publishers, Chelsea, MI, pp. 55–81.

Black, J.J., Baumann, P.C., 1990. Carecinogens and cancer in freshwater fish. Environ. Heallth Perspect. 90, 27–33.

Black, J., Fox, H., Black, P., Bock, F., 1985. Carcinogenic effects of river sediment extracts in fish and mice. In: Jolley, R.L. et al. (Eds), Water Chlorination: Chemistry, Environmental Impact and Health Effects, Vol. 5. Lewis Publishers, Chelsea, Michigan, pp. 415–427.

Bobra, A.M., Clark, A.G., Brown, J.R., 1978. Biochemical degradation of crude oil and oil components in pure and mixed culture systems. In: Afghan, B.K., Mackay, D. (Eds), Hydrocarbons and Halogenated Hydrocarbons in the Aquatic Environment. Plenum, New York, pp. 521–536.

Boehm, P.D., Quinn, J.G., 1976. The effects of dissolved organic matter in seawater on the uptake of mixed individual hydrocarbons and #2 fuel oil by a marine filter feeder bivalve Mercenaria mercenaria. Estuar. Coast. Marine Sci. 4, 93–105.

Boehm, P.D., Quinn, J.G., 1977. The persistence of chronically accumulated hydrocarbons in the hard shell clam Mercenaria mercenaria. Mar. Biol. 44, 227–233.

Bott, G.C., Crafford, S.D., 1985. Inhibition of phytoplankton photosynthesis by the water-soluble fraction of used lubricating oil. Mar. Poll. Bull. 16, 400–404.

Bott, T.L., Rogenmuser, K., Thorne, P., 1976. Effect of No. 2 fuel oil, Nigerian crude oil, and use of crankease oil on the metabolism of benthic algal communities, Sources, Effects and sinks of Hydrocarbon in the Aquatic Environment. American Institute of Biological Sciences, Washington DC,, pp. 374–393.

Brady, M.I., Al-Mujainy, I.S., Hernandez, M.D., 1993. Petroleum-derived hydrocarbons in water, sediment, and biota from Mina al-Fahl coastal waters. Mar. Poll. Bull. 26, 457–460.

Cairns, V.W., Fitzsimons, J.D., 1988a. The occurrence of epidermal papillomas and liver neoplasia in white suckers *(Catostomus commersoni)* from Lake Ontario. In: Niimi, A.J., Solomon, K.R. (Eds), Proceedings of the Fourteenth Annual Aquatic Toxicity Workshop, November 2-4, 1987, Toronto, Ontario, Canadian Technical Report of Fisheries and Aquatic Sciences, No. 1607, pp. 151–152.

Cairns, V.W., Fitzsimons, J.D., 1988b. The occurrence of epidermal papillomas and liver neoplasia in white suckers (*Catostomus commersoni*) from Lake Ontario. In: Niimi, A.J., Solomon, K.R. (Eds), Proceedings of the Fourteenth Annual Aquatic Toxicity Workshop, November 2-4, 1987, Toronto, Ontario, Canadian Technical Report of Fisheries and Aquatic Sciences, No. 1607, pp. 151–152.

Cajarville, M.P., Cancio, I., Ibaba, A., Orbea, A., 2003. Peroxisome proliferation as a biomarker in environmental pollution assessment. Microsc. Res. Tech. 61, 191–202.

Cancio, I., Cajaravillem, M.P., 2000. Cell biology of peroxisomes and their characteristics in aquatic organisms. Rev. Cytol. 199, 201–293.
Carls, M.G., Rice, S.D., Hose, J.E., 1999. Sensitivity of fish embryos to weathered crude oil: Part 1. Low-level exposure during incubation causes malformations, genetic damage, and mortality in larval pacific herring (Clupea pullasi). Environ. Toxicol. Chem. 18, 481–493.
Cavlieri, E.L., Rogan, E.G., 1992. The approach to understanding aromatic hydrocarbon carcinogenicity: the central role of the radical cation in metabolic activation. Pharmacol. Ther. 55, 83.
Chang, Y-J., Mathews, C., Mangold, K., Marlen, K., Hendricks, J., Bailey, G., 1991. Analysis of ras gene mutations in rainbow trout liver tumors initiated by aflatoxin B1. Mol. Carcinog. 4, 112–119.
Cohen, S.M., Klanning, J., Meek, M.E., Hill, R.N., Pastor, T., Lehman-McKeeman, L., Buchler, J., Longfellow, D.G., Seed, J., Dellacro, V., Fenner-Crisp, P., Patton, D., 2004. Evaluating the human relevance to chemically induced animal tumors. Toxicol. Sci. 78, 181–186.
Colborn, T.D., Dumanoski, D., Myers, J.P., 1997. Our Stolen Future. Penguin Books, New York.
Collier, T.K., Anulacion, B.F., Stein, J.E., Gloksoyr, A., Varanasi, U., 1995. A field evaluation of cytochrome P4501A as a biomarker of contaminant exposure in three species of flatfish. Environ. Toxicol. Chem. 14, 143–152.
Cooper, C.B., Doyle, M.E., Kipp, K., 1991. Risks of consumption of contaminated seafood: The Quicy Bay Case Study. Environ. Health Perspect. 90, 133–140.
Cormier, S.M., Racine, R.N., Smith, C.E., Dey, W.P., Peck, T.H., 1989. Hepatocellular carcinoma and fatty infiltration in the Atlantic tomcod, *Microgadus tomcod* (Walbaum). J. Fish Dis. 12, 105–116.
Couillard, C.M., 2002. A microscale test to measure petroleum oil toxicity to mummichog embryos. Environ. Toxicol. 17, 195–202.
Dahab, O.A., Al-Madia, H., 1993. Oil pollution in Qatar coastal sediments. Environ. Poll. 81, 113–116.
Dawe, C.J., Stanton, M.F., Schwartz, F.J., 1964. Hepatic neoplasms in native bottom-feeding fish of Deep Creek Lake. Maryland. Cancer Res. 24, 1194–1201.
DiSalvo, L.H., Guaw, H.E., Hunter, L., 1975. Tissue hydrocarbon burden of mussels as potential monitor of environmental hydrocarbon insult. Envir. Sci. Technol. 9, 247–251.
Douabul, A.A-Z., Al-Saad, H.T., Darmoian, S.A., 1984. Distribution of petroleum residues in surficial sediments from Shaat Al-Arab River and the Northwest region of the Arabian Gulf. Mar. Poll. Bull. 15, 198–200.
Douabul, A.A-Z., Abayachi, J.K., Al-Edamee, T.E., Ghani, A.A., Al-Saad, H.T., 1987. Polycyclic aromaticsin fish from the Arabian Gulf. Bull. Environ. Contamin. Toxicol. 38, 546–552.
Douabul, A. A-Z., Al-Shiwafi, N.A., 1998. Dissolved/dispersed hydrocarbons in the Arabian region. Mar. Poll. Bull. 36, 844–850.
Dunn, B.P., 1991. Carcinogenic adducts as indicators for the public health risks of consuming carcinogens. Environ. Health Perspect. 90, 11–116.
Eiskus, A., Monosson, E., McElroy, A.E., Stegeman, J.J., Woltering, D.S., 1999. Altered CYPIA expression in Fundulus heteroclitus adults and larvae: a sign of pollutant resistance. Aquat. Toxicol. 45, 99–112.
El-Deeb, K.Z., El-Ebiary, E.H., 1988. Total aromatic hydrocarbon content in the muscle and liver lipid extracts of two seabream fishes from the Arabian Gulf. J. Sci. Res. Agri. Biol. Sci. B6, 139–151.
Emara, H.I., 1990. Oil pollution in the southern Arabian Gulf and Gulf of Oman. Mar. Pollut. Bull. 21, 398–399.
Fabacher, D.L., Schmitt, C.J., Besser, J.M., Mac, M.J., 1988. Chemical characterization and mutagenic properties of polycyclic aromatic compounds in sediment from the tributaries of the Great Lakes. Envir. Toxicol. Chem. 7, 529–543.
Farley, C.A., Plutschak, D.L., Scott, R.F., 1991. Epizootiology and distribution of transmissible sarcoma in Maryland softshell clam Mya arenaria 1984–1988. Environ. Health Perspect. 90, 35–41.
Farrington, J.W., Davis, A.C., Frow, N.M., Rabin, K.S., 1982. No.2 fuel oil compounds in *Mytilus edulis*. Mar. Biol. 66, 15–26.
Fisher, W., Foss, S., 1993. A simple test for toxicity of number 2 fuel oil and oil dispersants to embryos of grass shrimp *Palaemonetes pugia*. Mar. Pollut. Bull. 267, 385–391.
Fitzsimmons, J.D., 1988. Hepatocellular adenoma in the slimy sculpin *Cottus cognatus*. J. Fish. Res. Bd. Can. 11, 89–91.

Fong, A.T., Dashwood, R.H., Cheng, R.I., 1993. Carcinogenicity, metabolism and Ki-ras proto-oncogene activation by 7,12-dimethylbenz[a]anthracene in rainbow trout embryos. Carcinogenesis 14, 629–635.

French-McCay, D., 2003. Development and application of damage assessment modeling: example assessment for the North Cape oil spill Mar. Poll. Bull. 47, 9–12.

Fucik, K.W., Carr, K.A., Balcom, B.J., 1995. In: Lane, P. (Ed.), Toxicity of Oil and Dispersed Oil to the Eggs and Larvae of Seven Marine Fish and Invertebrates from the Gulf of Mexico: The Use of Chemicals in Oil Spill Response. American Society for Testing and Materials, Philadelphia, PA, pp. 135–170, STP 1252.

Gardner, G.R., Pruell, R.J., 1988. A histopathological and chemical assessment of winter flounder, lobster and soft-shelled clam indigenous to Quincy Bay, Boston Harbor and an *in situ* evaluation of oysters including sediment (surface and cores) chemistry, Task II and III Report. US Environmental Protection Agency, Environmental Research Laboratory, Narragansett, RI.

Gardner, G.R., Pruell, R.J., Folmar, L.C., 1989. A comparison of both neoplastic and non-neoplastic disorders in winter flounder (*Pseudopleuronectes americanus*) from nine areas in New England, Programme and Abstracts: Fifth international symposium on responses of marine organisms to pollutants, 12–14 April 1989, Plymouth, UK, p. 123.

Grahl-Nielsen, J.T., Staveland, J.T., Wilhemson, S., 1978. Aromatic hydrocarbons in benthic organisms from coastal areas polluted by Iranian crude oil. J. Fish. Res. Bd. Can. 35, 615–623.

Grizzle, J.M., 1985. Black bullhead: an indicator of the presence of chemical carcinogens. In: Jolley, R.L et al. (Eds), Water chlorination: Chemistry, environmental impact and health effects, Vol. 5. Lewis Publishers, Chelsea, MI, pp. 451–462.

Guengerich, F.P.W., 1988. Roles of cytochrome P450 enzymes in chemical carcinogenesis and cancer chemotherapy. Cancer Res. 48, 2946–2954.

Guengerich, F.P.W., 1992. Metabolic activation of carcinogens. Pharm. Ther. 54, 17.

Gulec, I., Holdway, D.A., 1997. Toxicity of dispersant oil and dispersed oil to two marine organisms. In: Oil Spill Conference Proceedings, pp. 1010–1011.

Gulec, I., Holdway, D.A., 2000. Toxicity of crude oil and dispersed crude oil to gost shrimp *Palaemon serinus* and larvae of Australian bass *Macquaria novemaculaect*. Environ. Toxicol. 15, 91–98.

Hankinson, O., 1995. The aryl hydrocarbon receptor complex. Annu. Rev. Pharmacol. 35, 307.

Harshburger, J., Black, J.B., 1990. Epizootiology of neoplasms in bony fish of North America. Sci. Total Environ. 94, 1–32.

Hawkins, W.E., Walker, W.W., Lytle, J.S., Lytle, T.F., Oval, R.M., 1989. Comparative effect of 7, 12-dimethylbenzanthracene on the guppy (*Poecilia reticulata*). Aquat. Toxicol. 15, 63–82.

Hawkins, W.E., Walker, W.W., Overstreet, R.M., Lytle, J.S., Lytle, T.P., 1990. Carcinogenic effects of some polynuclear aromatic hydrocarbons on the Japanese medaka and guppy in waterborne exposures. Sci. Total Envir., 94, pp. 155–167.

Hayes, M.A., Smith, J.R., Rushmore, T.H., Crane, T.L., Thorn, C., Kocal, T.E., Ferguson, H.W., 1990. Pathogenesis of skin and liver neoplasms in white suckers from industrially polluted areas in Lake Ontario. Sci. Total Environ. 94, 105–123.

Hayes, M.O., Hoff, R., Michell, J., Scholz, D., Shigenaka, G., 1992. An introduction to coastal habitats and biological resources for oil spill response. Hazardous Materials Response and Assessment Division and National Oceanic and Atmospheric Admin., Report # HMRAD, 92–94.

Health Risk Assessment of Chemical Contaminants in Seafood, 1988. Final report. Tetra Tech, Inc. U.S. X-Office of Puget Sound, Seattle, WA. Environmental Protocol 91019-88-249. Conference Center, Leesburg, VA.

Heintz, R.A., Short, J.W., Rice, S.D., 1999. Sensitivity of fish embryos to weathered crude oil: Part II. Increased mortality of pink salmon (*Oncorhynchus gobusha*) embryos incubating downstream from weathered Exxon Valdez crude oil. Environ. Toxicol. Chem. 18, 494–503.

Hendricks, J.D., Meyers, T.R., Shelton, D.W., 1984. Histological progression of hepatic neoplasia in rainbow trout (*Salmo gairdneri*). Natl Cancer Inst. Monogr. Ser. 65, 321–336.

Hendricks, J.D., Meyers, T.R., Shelton, D.W., Casteel, J.L., Bailey, G.S., 1985. Hepatocarcinogenicity of benzo(a)pyrene to rainbow trout by dietary exposure and intraperitoneal injection. J. Natl Cancer Inst. 74, 839–851.

Hendricks, J.D., Cheng, R., Shelton, D.W., Pereira, C.B., Bailey, G.S., 1994. Pendent carcinogenicity and frequent Ki-ras proto-oncogene activation by dietary *N*-nitrosodiethylamine in rainbow trout. Fundam. Appl. Toxicol. 23, 53–62.

Hileman, B., 1988. The Great Lakes clean-up effort, Chemical and Engineering News, February , 8. 22–27.

Hinton, D.E., Klaunig, J.E., Jack, R.M., Lipsky, M.M., Trump, B.F., 1981. *In vitro* evaluation of the channel catfish *Ictalurus punctatus* (Rafinesque) as a test species in chemical carcinogenesis studies. In: Branson, D.R., Dickson, K.L. (Eds), Aquatic Toxicology and Hazard Assessment. American Society for Testing and Materials, PA, pp. 226–238.

Hose, J., Hannah, J., Landohlt, M., Miller, B., Felton, S., Iwoska, W., 1981. Uptake of benzo(a)pyrene by gonadal tissue of flatfish (*Family Leuronectidae*) and its effects on subsequent egg development. J. Toxicol. Environ. Health 7, 999–1000.

Hu, Z., Wells, P.G., 2004. Human interindividual variation in lymphocyte UDP-Glucuronyltransferases as a determinat of *in vitro* benzo(a)pyrene covalent binding and cytotoxicity. Toxicol. Sci. 78, 30–32.

Huff, J., 2003. IARC and the DEHPguagmire. J. Occup. Environ. Health 9, 402–414.

Hyland, J.L., Schneider, E.D., 1976. Petroleum hydrocarbons and the effects on marine organism, populations, communities, and ecosystems. In: Sources, Effects, and Sinks of Hydrocarbons in Aquatic Environment. Am. Inst. Biol. Sci.

IARC, 1983. International Agency for Research in Cancer. The Monographs. The Evaluation of the Carcinogenic Risks of Chemicals to Humans. Polynuclear Aromatic Hydrocarbons. Part 1. Vol. 32, 35–451, and Vol 38, 309-314. IARC, Lyon, France.

James, M.O., Fouts, J.R., Bend, J.R., 1976. Xenobiotic metabolism enzymes in marine fish. In: Khan, M.A.Q. (Ed.), Pesticides in Aquatic Environments. Plenum Publications, New York, pp. 171–189.

James, M.O., Khan, M.A.Q., Bend, J.R., 1979. Hepatic microsomal mixed-function oxidase activities in several marine species common to coastal Florida. Compar. Biochem. Physiol. 62C, 155.

James, M.O., Heard, C.S., Hawkins, W.E., 1988. Effect of 3-methylcholanthrene on monooxygenase, epoxide hydrolase, and glutathione S-transferase activities in small estuarine and freshwater fish. Aquat. Toxicol. 12, 1–15.

Johnson, L., Stein, J., Collier, T., Sol, S., Varanasi, U., 1995. Effect of exposure to Purdohoe Bay crude oil on reproductive function in gravid female fish. Environ. Sci. 3, 67–81.

Khan, M.A.Q., 1980. Induction of Drug metabolizing enzymes by insecticides and other xenobiotics. Pharmacol. Therap. 11, 43.

Khan, R., Kiceniuk, J., 1984. Histopathological effects of crude oil on Atlantic tomcod following chronic exposure. Can. J. Zool. 62, 2038–2043.

Khan, M.A.Q., Al Ghais, S.M., Almarri, S., 1995. Petroleum hydrocarbons in fish from coastal U.A.E. Arch. Environ. Contamin. Toxicol. 29, 517.

Khan, M.A.Q., Qadri, S.Y., Tomar, S., Fish, D., Gururajan, L., Poria, M.S., 1998. Induction of cytochrome P-4502B in semiaquatic frog Rana pipiens. Biocem. Biophys. Res. Commun. 244, 737.

Kocan, R.M., Landolt, M.L., 1984. Alterations in patterns of excretion and other metabolic functions in developing fish embryos exposed to benzo(a)pyrene. Helgolander Meeresunters. 37, 493–504.

Kocan, R.M., Sabo, K.M., Landolt, M.L., 1985. Cytotoxicity/genotoxicity: the application of cell culture techniques to the measurement of marine sediment pollution. Aquat. Toxicol. 6, 165–177.

Kocan, R.M., Westernhagen, H.V., Landolt, M.L., Furstenberg, G., 1987. Toxicity of sea-surface microlayer: II. Effects of hexane extract on Baltic herring (*Clupea harengus*) and Atlantic cod (*Gadus morhua*) embryos. Marine Environ. Res. 23, 291–305.

Kohler, A., Wahl, E., Soffker, K., 2002. Functional and morphological changes of lysosomes as prognostic biomarkers of toxic liver injury in a marine flat fish (*Pleuronectes vetulus*). Environ. Toxicol. Chem. 21, 2434–2444.

Krahn, M.M., Meyers, M.S., Burrows, D.G., Malins, D.C., 1984. Determination of metabolites of xenobiotics in bile of fish from polluted waterways. Xenobiotica 14, 633–646.

Krahn, M.M., Burrows, D.G., Macleod, W.D., Malins, D.C., 1987. Determinations of aromatic compounds in hydrolyzed bile of English sole (*Parophrys vetulus*) from polluted sites in Puget Sound. Washington Arch. Environ. Contam. Toxicol. 16, 511–522.

Laramaira, P., Livingston, D.R.L., 1997. Armoatic hydrocarbon–quinone-mediated reactive oxygen species in hepatic microsomes of the fish *Platichthys flessus* L. Compar. Biochem. Physiol. 117C, 131.

Lee, R.F., 1977. Accumulation and turn over of petroleum hydrocarbons in marine organisms. In: Wolfe, D.A. (Ed.), Fate and Effects of Petroleum hydrocarbons in marine Ecosystem and Organsims. Pergammon, New York, pp. 60–70.

Lee, R.F., 1981. Effects of crude oil on growth and mixed function oxygenase activity in polychaete, Neries spp. In: Biological Monitoring of Marine Pollutants. Academic Press, New York, pp. 325–334.

Lee, R.F., Stotzenbach, J., Singer, S., Tenore, K.R., 1980. Effects of crude oil on growth and mixed function oxidase activity in Neries spp. In: Biological Monitoring of Marine Pollution. Academic Press, New York, pp. 323–334.

Lee, L.E.J., Manu, T., Pagniello, K., Mikaelian, I., DeLaroutanne, Y.H., 2001. Examination of EROD activity and fibronectin levels in lake whitefish as biomarker of neoplasms. In: Greenberg, B.M., Hall, R.N., Roberts, M.H. Jr., Gensemer, R.W. (Eds), Environmental Toxicology & Risk Assessment, Vol. 10, pp. 193–210, STP., ASTM.

Lotlikar, P.D., Thee, E.C., Insetta, S.M., Clearfield, M.S., 1984. Modulation of microsome mediated aflatoxin B, binding to exogenous and endogenous DNA by cytosolic glutathione S-transferases in rat and hamster livers. Carcinogenesis 5, 269–276.

Lotufo, G.R., 1997. Toxicity of sediment associated PAHs to an estuarine copepod, effects on survival, feeding, reproduction and behavior. Mar. Environ. Res. 44, 149–166.

Lu, X.X., Reible, D.D., Fleeger, J.W., 2004. Bioavailability and assimilation of sediment-associated benzo (a) pyrene by Ilyodrilus templetoni (Oligochaeta). Environ. Toxicol. Chem. 23, 57–64.

Maccubbin, A.E., Chidambaram, S., Black, J.J., 1988. Metabolites of aromatic hydrocarbons in the bile of brown bullheads Ictalurus nebulosus. J. Great Lakes Res. 14, 101–108.

Maccubbin, A.E., Iljima, H., Ersuing, N., Davrrudzik, J.B., Petrzyc, H.B., Wallace, J.C., Budzinski, E.E., Freund, H.G., Box, H.C., 2000. Double-based lesions in DNA by free radicals. Arch. Biochem. Biophys. 375, 119–123.

Malan, D.E., 1992. The effects of Qatar light crude oil on the salt marsh crab Sesarma calenecta in the Swartkeps estuary, Port Elizabeth, South Africa. M.Sc. Thesis, University of Port Elizabeth, South Africa.

Malins, D.C., Haimanot, R., 1990. Amino-5-formamidopyrimidine, 8-hydroxyguanine and 8-hydroxy-adenine in DNA from neoplastic liver of English sole exposed to carcinogens. Biochem. Biophys. Res. Commun. 173, 614–619.

Malins, D.C., McCain, B.B., Brown, D.W., Myers, M.S., Krahn, M.M., Chan, S., 1987a. Toxic chemicals, including aromatic and chlorinated hydrocarbons and their derivatives, and liver lesions in white croaker (*Genyonemus lineatus*) from the vicinity of Los Angeles. Environ. Sci. Technol. 21, 765–770.

Marty, G.D., Short, J.W., Dambach, D.M., Willits, N.H., Heintz, R.A., Rice, S.D., Stegeman, J.J., Hinton, D.E., 1997. Ascites, premature emergence increased gonadal cell apoptosis and cytochrome P4501A induction in pink salmon larvae continuously exposed to oil-contaminated gravel during development. Can. J. Zool. 75, 989–1007.

May, E.B., Lukacuvic, R., King, H., Lipsky, M.M., 1987. Histopathology and neoplastic alterations in liver of white perch (*Morone americanus*) from the Chesapeak Bay. J. Natl Cancer Inst. 79, 137–143.

McCain, M.S., Myers, D.W., Brown, M.M., Krahn, W.T., Roubal, M.H., Schiewe, J.T., Landahl, J.T., Chan, S-L., 1987. Field and laboratory studies of the etiology of liver neoplasms in marine fish from Puget Sound. Environ. Health Perspect. 71, 5–16.

McCann, J.E., Choi, E., Yamasaki, E., Ames, B., 1975. Detection of carcinogens as mutagens in the Salmonella/microsome test: assay of 300 chemicals. Proc. Nat. Acad. Sci. USA. 72, 5135–5139.

McGurk, M.D., Brown, E.D., 1996. Egg-larval mortality of Pacific herring in Prince William Sound, Alaska, after the Exxon Valdez oil spill. Can. J. Fish Aquat. Sci. 53, 2343–2354.

McMahon, G., Huber, U., Stegeman, J.J., Wogan, G.N., 1988. Identification of a c-Ki-ras oncogene in a neoplasm isolated from winter flounder. Mar. Environ. Res. 24, 345–350.

McMahon, G., Huber, U., Moore, M.J., Stegeman, J.J., Wogan, G.N., 1990. Mutations in c-Ki-ras oncogenes in diseased livers of flounder from Boston Harbor. Proc. Natl Acad. Sci. USA 87, 841–845.

Metcalfe, C.D., 1989. Tests for predicting carcinogenicity in fish. CRC Crit. Rev. Aquat. Sci. 1, 111–129.

Metcalfe, C.D., Sonstegard, R., 1984. Microinjection of carcinogens into rainbow trout embryos: an *in vivo* carcinogenesis assay. J. Natl Cancer Inst. 73, 1125–1131.

Metcalfe, C.D., Cairns, V.W., Fitzsimons, J.D., 1988. Microinjection of rainbow trout at the sac-fry stage: a modified trout carcinogenesis assay. Aquat. Toxicol. 14, 347–353.

Moore, J.D., Elston, R.A., Drum, A.S., Wilkinson, M.T., 1991. Alternate pathogenesis of systemic neoplasia in bivalve mollusk Mytilus. J. Invertebr. Pathol. 58, 231–243.

Murchelano, R.A., Wolke, R.E., 1985. Epizootic carcinoma in the winter flounder (*Pseudopleuronectes americanus*). Science 228, 587–589.

Murchelano, R.A., Wolke, R.E., 1991. Neoplasms and neoplastic liver lesions in white flounder, Pseudo-pleuronectes americanus, from Boston Harbor, Massachussetts. Environ. Health Persp. 90, 17.

Myers, M.S., Rhodes, L.D., 1988. Morphological similarities and parallels in geographic distribution of susceptible toxicopathic liver lesions in rock sole (*Lepidopsetta bilineata*), starry flounder (*Platichthys stellatus*), Pacific staghorn sculpin (*Leptocottus armatus*) from urban and nonurban embayments in Puget Sound, Washington. Aquat. Toxicol. 11, 410–411.

Myers, M.S., Landahl, J.L., Krahn, M.M., Johnson, L.L., McCain, B.B., 1990. Overview of studies on liver carcinogenesis in English sole from Puget Sound. Evidence for a xenobiotic chemical etiology. I. Pathology and epizootiology. Sci. Total Environ. 94, 33–50.

Myers, M.S., Stehr, C.M., Olson, O.P., 1991a. Relationships between toxicopathic hepatic lesions and exposure to chemical contaminants in English sole (*Pleuronectes vetulus*), starry flounder (*Platichthys stellatus*), and white croaker (*Genyonemus lineatus*) from selected marine sites on the Pacific Coast, USA. Environ Health Perspect 102, 200–215.

Myers, M.S., Landahl, J.T., Krachu, M.M., McCain, R.B., 1991b. Relationship between hepatic neoplasms and related lesions and exposure to toxic chemicals in marine fish of U.S. West Coast. Environ. Health Perspect. 90, 7–15.

National Research Council, 1985. Oil in the Sea, Inputs, Fates, and Effects. Washington, National Academy Press, p. 601.

Neff, J.M., Cox, B.A., Dixet, D., Anderson, J.W., 1976. Accumulation and release of petroleum derived hydrocarbons by four species of marine animals. Mar. Poll. Res. 38, 279–289.

Nishimoto, M., 1986. Metabolism of Benzo(a)pyrene by English Sole, Starry Flounder, and Rat. Doctoral dissertation. University of Washington, Seattle, Washington.

Nishimoto, M., Varanasi, U., 1985. Benzo(a)pyrene metabolism and DNA adduct formation mediated by English sole liver enzymes. Biochem. Pharmacol. 34, 263–268.

Nishimoto, M., Varanasi, U., 1986. Metabolism and DNA adduct formation of benzo(a)pyrene and the 7,8-dihydrodiol of benzo(a)pyrene by fish liver enzymes. In: Cooke, M.C., Dennis, A.J. (Eds), Polynuclear Aromatic Hydrocarbons: Fate and Effects. Battelle Press, Columbus, OH, pp. 685–699.

O'Brien, P.Y., Dixon, P.S., 1976. The effects of oils and oil components on algae: a review. Br. Phycol. J. 11, 115–142.

Omer, G.A., Mathews, C., Hendricks, J.D., Carpenter, H.M., Bailey, G.S., Williams, D.E., 1995. Dehydroepiandrosterone is a complete hepato- N carcinogen and potent tumor promoter in the absence of peroxisome proliferation in rainbow trout. Carcinogenesis 16, 2893–2898.

Osborne, M.R., Crosby, N.T., 1987. Benzopyrenes. Cambridge University Press, Cambridge.

Ostrander, G.K., Landolt, M.L., Kocan, R.M., 1988. The ontogeny of coho salmon (*Onco-rhynchus kisutch*) behavior following embryonic exposure to benzo(a)pyrene. Aquat. Toxicol. 13, 325–346.

Ostrander, G.K., Landolt, M.L., Kocan, R.M., 1989. Whole life history studies of coho salmon (*Ooniskus kisutch*) following embryonic exposure to benzo(a)pyrene. Aquat. Toxicol. 15, 109–126.

Ostrander, G.K., Anderson, J.J., Fisher, J.P., Landolt, M.L., Kocan, R.M., 1990. Decreased performance of rainbow trout *Oncorhynchus mykiss* emergence behaviors following embryonic exposure to benzo(a)pyrene. Fish Bull. 88, 551–555.

Overstreet, R.M., 1988. Aquatic pollution problems, Southeastern U.S. coasts: histopathological indicators. In: Malins, D.C., Jensen, A., Moore, M.N. (Eds), Aquatic Toxicology, Vol. 11. Elsevier Science Publishers BV, Amsterdam, pp. 213–239.

Papadopoulou, D., Scamara, C., 2002. Polycyclic aromatic hydrocarbon contamination and Lumistox solvent extracts toxicity of marine sediments in the north Aegian Sea. Greece Environ. Toxicol. 17, 556–566.

Payne, J.F., 1976. Field evaluation of benzo(a)pyrene hydroxylase induction as a monitor for marine petroleum pollution. Science 191, 945.

Payne, J.F., May, N., 1979. Further studies on the effects of petroleum hydrocarbons on mixed-function oxidase in marine organisms. In: Khan, M.A.Q., Lech, J.J., Menn, J.J. (Eds), Pesticide and Xenobiotic Metabolism in Aquatic Organisms. American Chemical Society, Washington, DC, pp. 339–348.

Peck-Miller, K.A., Myers, M., Collier, T.K., Stein, J.E., 1988. Complete cDNA sequence of the kiras protooncogene in the liver of wild English sole (*Pleuronectes vetulus*) and mutation analysis of hepatic neoplasms and other toxicopathic lesions. Mol. Carcinog. 23, 202–216.

Reichert, W.L., Myers, M.S., Peck-Miller, K., French, B., Anulacea, B.F., Collier, T.K., Stein, J.E., Varanasi, U., 1998. Molecular epizootiology of genotoxic events in marine fish linking contaminant exposure, DNA damage and timor level alterations. Mut. Res. 411, 215–225.

Rice, C.L., Putnam, D.J., Plewa, F.R., 1986. A preliminary survey of tumors in brown bullheads in Presque Isle Bay, Lake Erie, Erie, Pennsylvania. Department of the Interior, U.S. Fish and Wildlife Service, State College Field Office, Resource Contaminant Assessment Report No. 86-1, State College, PA.

Rice, S.D., Short, J.W., Karinen, J.F., 1976. Toxicity of Cook Inlet crude oil and No.2 fuel oil to several Alaskan marine fishes and invertebrates. In: ERDA, EPA, ELM, and API, Sources, Effects, and Sinks of Hydrocarbons in the Aquatic Environment. Amer. Inst. Biol. Sciences, Washington, DC, pp. 395–406.

Shin, K-J., Kim, S-H., Kim, D., Kim, Y-H., Lee, H-W., Chang, Y-S., Gu, M-B., Ryu, S-H., Suh, P-G., 2004. 2,2′,4,6,6′-Pentachlorobiphenyl induces mitotic arrest and p53 activation. Toxicol. Sci. 78, 215–221.

Shriadah, M.A., 2000. Levels and distribution of polycyclic aromatic hydrocarbons in the coastal waters and sediments of the United Arab Emirates in the Arabian Gulf and Gulf of Oman. Water Air Soil Poll. 119, 247–256.

Shriadah, M.A., 2001. Petroleum hydrocarbon contamination in Arabian Gulf fish tissue. Bull. Environ. Contamin. Toxicol. 67, 560–567.

Singer, M.M., Smallheer, D.L., Tjeerdema, R.S., Martin, M., 1991. Environ. Toxicol. Chem. 10, 1367–1374.

Smith, C.E., Peck, T.H., Klauda, R.J., McLaren, J.B., 1979. Hepatomas in Atlantic tomcod *Microgadus tomcod* (Walbaum) collected in the Hudson River estuary in New York. J. Fish Dis. 2, 313–319.

Sonstegard, R.A., Leatherland, J.F., 1984. Comparative epidemiology: the use of fishes in assessing carcinogenic contaminants. In: Cairns, V. et al. (Eds), Contaminant Effects on Fisheries, Advances in Environmental Science and Technology, Vol. 16, pp. 223–232.

Steering Committee, 1991. Chemically-contaminated aquatic food resources and human cancer risk: retrospect. Environ. Health Perspect. 90, 149–154.

Steering Committee, 1995. The Use of Chemical Counter Measures. Product Data for Oil Spills. Planning & Response Corporation. U.S. Coastal guard. National Oceanic and Atmospheric Administration Workshop Proceedings. Xerox Document University.

Stegeman, J.J., 1989. Cytochrome P450 forms in fish: catalysis, immunology, and sequence similarities. Xenobiotica 8, 1093–1110.

Stegeman, J.J., Lech, J.J., 1991. Cytochrome P450 monoxygenase system in aquatic species: carcinogen metabolism and biomarkers for carcinogen and pollutant exposure. Environ. Health Perspect. 90, 101–119.

Stegman, J.J., Teal, J.M., 1973. accumulation, retention, and release of petroleum hydrocarbons by the oyster *Crassostrea virginica*. Mar. Poll. 22, 37–44.

Stein, J.E., Hom, T., Casillas, E., Friedman, A., Varanasi, U., 1987. Simultaneous exposure of English sole (Parophrys vetulus) to sediment-associated xenobiotics: part II. Chronic exposure to an urban estuarine sediment with added [3H]benzo(a)pyrene and [^{14}C]polychlorinated biphenyls. Mar. Environ. Res. 22, 123–149.

Stein, J.E., Reichert, W.L., French, B., Veranansi, U., 1993. ^{32}P-Postlabeling analysis of DNA adduct formation and persistence in English sole (pleuronecies vetulus) exposed to benzo (a) pyrene and 7H-dibenzo(c,g) Carbazole. Chem. Biol. Interact. 88, 55–69.

Stroker, S., Seager, S.L., 1976. Environmental Chemistry: Air and Water Pollution. Scott, Freeman, and Co., Glenview, IL, p. 231.

Tomatis, L., 2002. The IARC monograph programs: changin attitudes towards public health. Int. J. Occup. Environ. Health 8, 144–152.

Torten, M., Liu, Z., Okihiro, M.S., Teh, Si., Hinton, D.E., 1996. Action of ras oncogene mutations and hepatocarcinogenesis in medaka (*Oryzias latipes*) exposed to diethylnitrosamine. Mar. Environ. Res. 42, 93–98.

Trescot, C., Sumper, J., Bromage, N., 1988. *In vitro* oxygen uptake and processing of vitellogenin in the rainbow trout salmo gairdneri. J. Exp. Zool. 246, 171–179.

U.S. Environmental Protection Agency, 1995. The Effects of Great Lakes Contaminants on Human Health, Report to Congress. EPA 905-R-95-107, September 1995. Great Lakes National Program Office, Chicago, IL.

Varanasi, U., Nishimoto, M., Stover, J., 1983. Analyses of biliary conjugates and hepatic DNA binding in benzo(a)pyrene-exposed English sole. In: Cooke, M.W., Dennis, A.J. (Eds), Polynuclear Aromatic Hydrocarbons: Eighth International Symposium on Mechanisms, Methods and Metabolism. Battelle Press, Columbus, OH, pp. 1315–1328.

Varanasi, U., Nishimoto, M., Reichert, W.L., Eberhart, B-T.L., 1986. Comparative metabolism of benzo(a)pyrene and covalent binding to hepatic DNA in English sole, starry flounder, and rat. Cancer Res. 46, 3817–3824.

Varanasi, U., Stein, J.E., Nishimoto, M., 1989a. Biotransformation and disposition of polycyclic hydrocarbons in fish. In: Varanasi, U. (Ed.), Metabolism of Polycyclic Aromatic Hydrocarbons in the Aquatic Environment. CRC Press, Boca Raton, FL, pp. 93–149.

Varanasi, U., Reichert, W.L., Eberhart, B-T.L., Stein, J.E., 1989b. Formation and persistence of benzo(a)pyrene-diolepoxide-DNA adducts in liver of English sole (*Parophrys vetulus*). Chem. Biol. Interact. 69, 203–216.

Villeneuve, D.L., Khim, J.S., Kannan, K., Giesy, J.P., 2002. Relative potencies of polycyclic aromatic hydrocarbons to induce dioxin-like and esterogenic responses in three cell lines. Environ. Toxicol. 17, 128–137.

Vogelbein, W.K., Fournie, J.W., van Veld, P.A., Hugget, R.J., 1990. Hepatic neoplasms in the mummighog *Fundulus heteroclitus* from a creosote contaminated site. Cancer Res. 50, 5978.

Von Westernhagen, H., 1988. Sublethal effects of pollutants on fish eggs and larvae. In: Hoar, W., Randall, D. (Eds), Fish Physiology. Academic Press, San Diego, CA, pp. 253–345.

Von Westernhagen, H., Landolt, M., Kocan, R., Furstenberg, G., Janssen, D., Kremling, K., 1987. Toxicity of sea-surface microlayer: I. Effects on herring and turbot embryos. Mar. Environ. Res. 23, 273–290.

Whipple, J., Yocom, T., Smat, D., Cohen, M., 1978. Effects of chronic concentrations of petroleum hydrocarbons on gonadal maturation of starry flounder (*Pleuronectes vetulus*). Proceed. of the Conference on Assessment of Oil Spills., Keystone, CO.

White, P.A., Robitaille, S., Rasmussen, J.B., 1999. Heritable reproductive effects of benzo(a)pyrene on the fathead minnoe (*Pimephales promelas*). Environ. Toxicol. Chem. 18, 1843–1847.

Widdows, J., Bakke, T., Bayne, B.L., Doukin, P., Livingston, D.R., Lowe, D.M., Monroe, M.N., Evans, S.V., Moore, S.L., 1982. Responses of *Mytilus edulis* on exposure to the water-accumulated fraction of North sea. Mar. Biol. 67, 15–31.

Wilson, C.D., 2004. Molecule switch circuits in toxicology: a dimmer switch for dioxin. Toxicol. Sci. 78, 178–180.

Wirgin, I., Currie, D., Garte, S.J., 1989. Activation of the K-ras oncogene in liver tumors of Hudson River tomcod. Carcinogenesis 10, 2311–2315.

Wirgin, I.I., Kreamer, G.L., Garte, S.J., 1991. Genetic polymorphism of cytochrome P-450IA in cancer-prone Hudson River tomcod. Aquat. Toxicol. 19, 205.

Woodhead, R.J., Law, R.J., Mathiesen, P., 1999. Polycyclic aromatic hydrocarbons in surface sediments around England and Wales and their possible biological significance. Mar. Poll. Bull. 38, 773–790.

Oil Pollution and its Environmental Impact in the Arabian Gulf Region
M. Al-Azab, W. El-Shorbagy and S. Al-Ghais, editors

Chapter 13

Assessment of biological characteristics on coastal environment of Dubai during oil spill (14 April 2001)

Shahid Mustafa* and M.A.H. Deshgooni

Environment Protection and Safety Section, Dubai Municipality, Dubai, UAE

Abstract

Georgian flagged tanker carrying 1300 metric tons of fuel oil sank 16.5 nautical miles north of Jabal Ali Port, Dubai in the ROPEME Sea on 14th April 2001. The 11-man crew, including the Captain, was rescued by the United Arab Emirates Coast Guard. Abnormal concentrations of Total Petroleum Hydrocarbons in the water column exposed pelagic biota to high levels of oil residues. Level of chlorophyll-a varied widely and was unevenly distributed. Phytoplankton species like Rhizosolenia stiliformis, R. imbricata, R. stolterfothi, Nitzschia, Prorocentrum micans, Coscinodiscus and Leptocylindrus danicus were damaged with coating of black–brown layer of petroleum hydrocarbons around as well as inside the cell. Zooplankton were also severely affected and damage clearly demarcated in the species of Copepods, fish eggs, fish larvae, Segitta sp. and Lucifer sp. The Zooplankton species showing the oil coasting on body surface and ingestion of hydrocarbons in the body leading to death in extreme cases. Heavy impact has been noticed in the zooplankton and drastic reduction in the population density of copepods was noticed. As the petroleum hydrocarbons had not sedimented, no adverse impact on macro-benthos was evident. Also no evidence of damage was seen in the fish and other marine life. Monitoring of the same area during 2002 showed normal water quality and biological characteristics indicating the complete recovery of the marine environment.

1. Introduction

Oil reserves of the Gulf and the surrounding landmass are the largest in the world and represent about 655 of the total world reserve. The estimated oil reserves of the Gulf are of tremendous economic and strategic importance to the Gulf and the world (Hinrichsen, 1996). On the annual basis, around 0.6 billion metric tons of oil or about 60% of the world's requirement, is carried by some 25,000 tankers from 34 major oil terminals of the Gulf to various parts of the world. It has been estimated that during peak period one ship passes the Strait of Hormuz every 6 min (Reynolds, 1993). The heavy tanker traffic through the narrow Strait of Hormuz raises serious concerns about collisions causing oil spills.

*Corresponding author. *E-mail address:* vision12@emirates.net.ae (S. Mustafa).

In 1994 alone approximately 15 collisions involving oil tankers were reported near the shores of the United Arab Emirates (Mahmoudi, 1997).

In the marine environment the spilled oil goes through a variety of transformations involving physical, chemical, and biological processes. Surface oil, tar balls, and floating oil are frequently seen through the ROPME Sea Area (ROPME, 1999). Reliable scientific data on background levels of oil and oil-related compounds in various compartments of the Gulf ecosystem are sparse. Some efforts were made in 1980s to monitor and establish background levels of oil and non-oil pollutant in the Gulf (Fowler, 1985; Literathy et al., 1986; KFUPM/RI, 1987). However, a synoptic picture of the distribution of Petroleum Hydrocarbons (PHC) in the Gulf did not emerge until the results of post-Gulf war cruises were published (Al-Abdali et al., 1986; Al–Ghadban et al., 1996; Massoud et al., 1998; Otsuki et al., 1998). Information on biological characteristics assessment during the oil spill is sparse. In view of this aspect a case study was undertaken during oil spill off Dubai on 16th and 17th April 2001. The possible impact of the spill on phytoplankton, zooplankton were studied.

1.1. Background of spill

Tanker *Zainab* carrying smuggled oil from Iraq sank near Jabal Ali on 14th April 2001. The ship had been intercepted several days earlier for violating UN sanctions against Iraq. "The ship was en route to a holding area in international waters for sanction-busting ship when it sank."

The tanker which started sinking at around 2 p.m. on 14th April 2001 was completely sank by the evening of 14th April 2001 and spilled 500 metric tons of its cargo, threatening the coastal marine environment of Dubai. Later due to prevailing environmental conditions and use of dispersants, the spill dispersed down into several patches of varying sizes and scattered in the region.

Aerial surveys conducted during 15th to 18th April 2001 indicate the large number of brownish black patches of degraded oil scattered throughout the coastal waters off Dubai. Due to high winds of shamals on 16th and 17th April 2001, the oil spill reached the Jumeirah and Mamzar coast on 19th April 2001.

2. Materials and methods

Measurement of temperature, pH and salinity were carried out *in situ* by the pre-calibrated Hydrolab. Sample of nutrients analysis were collected in 1 l of clean polyethylene bottle and analyzed as per the APHA (1995) procedure for the analysis of nitrate and phosphate. Samples of petroleum hydrocarbons were collected below 1-m surface depth and analyzed according to Manual of Oceanographic Observation and Pollutant Analysis Method (MOOPAM, 1999) as chrysene fraction for the petroleum hydrocarbons. Total Petroleum Hydrocarbons (TPH) were analyzed during the oil spill according to APHA (1995) procedure 552°C. Estimation of chlorophyll-a was carried out according to manual of biological parameters (Strickland and Parsons, 1972). Duplicate samples for phytoplankton were collected in 500 ml polyethylene bottles. They were fixed in lugols iodine solution and preserved with formaline. Enumeration and identification of phytoplankton were done under Sedgwick Rafter chamber having a capacity of 1 ml. A portion of sample (25–50%) was analyzed under the microscope for faunal composition and population count. The population was estimated as number of organisms in 100 m^3 water and biomass on volume basis.

3. Results and discussion

The data of oil spill have been collected on 16th and 17th April 2001 from the six monitoring stations as demarcated in Figure 1. The baseline data collected during 2000 and post-spill monitoring data obtained during 2002 were used for the comparison of water quality and biological characteristics. These data have been collected from the routine monitoring stations from the Dubai Municipality Marine Monitoring Program (Fig. 1).

It is evident from the baseline results that the coastal environment of Dubai are free from any sort of environmental stress even in the presence of industrialization along Jabal Ali area. The post-monitoring indicates insignificant impact of oil spill on the marine environment.

No changes in the temperature, pH and salinity variation have been seen during the course of oil spill; the high values of temperature and salinity during 2000 and 2002 at DEWA and DUBAL are due to the desalination discharges from the Power Generation and Desalination Plants located along these location (Table 1). Levels of nutrients (nitrate and phosphate) show slight variations with the normal baseline and post-monitoring values (Table 2).

PHC was analyzed as TPH during oil spill monitoring. In all the samples the levels of TPH were significantly recorded as high as 20,000 times than the normal value. Mostly the significant impacts were seen on 17th April 2001 at stations 1 and 2, when the oil residues dissolved in the water column perhaps due to the use of dispersant. Although the values recorded on 16th October 2001 at stations 3, 5, and 6 were also very high; however comparable with higher values reported by El-Samara et al. (1986) from Saudi Arabia (546 $\mu g\ l^{-1}$) and Qatar (428 $\mu g\ l^{-1}$) from the highly localized area of oil pollution.

Levels of petroleum hydrocarbons along Dubai coastline are generally very lower during 2000 (0.3–1.06 $\mu g\ l^{-1}$) and 2002 (0.9–5.2 $\mu g\ l^{-1}$) and very well comparable with the petroleum hydrocarbons levels (0.48–16.8 $\mu g\ l^{-1}$) of South-Western Region of United Arab Emirates (Fowler, 1985). The results also indicated that the mean levels of post-spill monitoring during 2002 have higher levels (2.6 $\mu g\ l^{-1}$) than the baseline data of 2000 (0.6 $\mu g\ l^{-1}$), this could be attributed to the increasing oil tanker traffic, minor accidental oil spills, and other oil related activities in Arabian Gulf.

In context with biological characteristics, the data indicates significant impact on pelagic plankton organisms. Prevailing biological characteristics of Dubai were evaluated based on the qualitative and quantitative data on organisms from the pelagic levels including phytoplankton and zooplankton. Capuzzo (1987) reported that the biological effects of PHC on marine organism are a function of factors like persistence and bio-availability of the hydrocarbon, ability of the organisms to accumulate and metabolize them, and the amount of contaminants interfering with normal metabolisms.

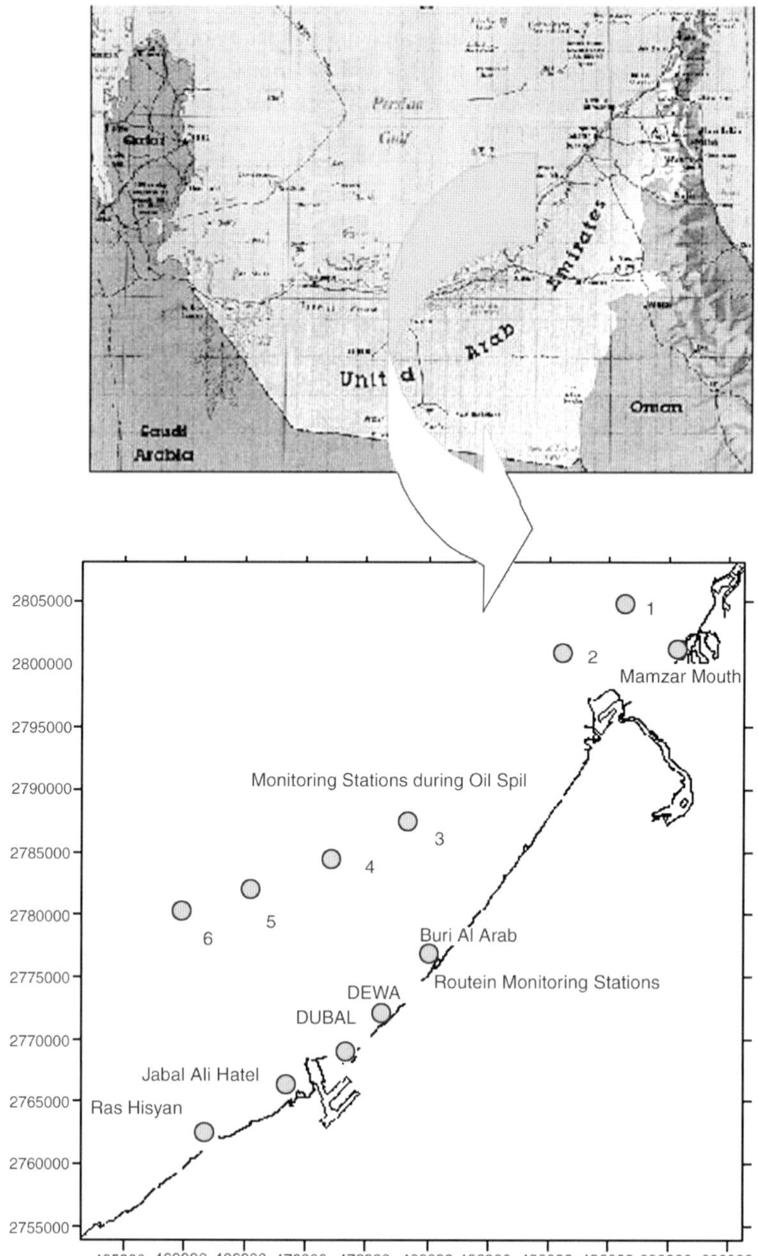

Figure 1. Oil spill and routine monitoring stations along Dubai coastline.

Table 1. Water quality along the coastal environment of Dubai during oil spill compared with baseline and post-spill monitoring results (average in parenthesis).

Parameters	Off Mamzar	Off Burj Al Arab	Off DEWA	Off DUBAL	Off Jabal Ali	Off Ras Hisyan
Temperature (°C)						
Baseline data (2000)	21–34	20–35	21–43	20–43	19–33	20–34
	(24)	(23)	(23)	(24)	(24)	(23)
Oil spill monitoring (2001)[1]	27–33	26–30	26–30	27–30	24–29	25–30
	(28)	(28)	(26)	(25)	(27)	(25)
Post-spill monitoring (2002)	20–34	19–35	19–34	20–34	19–34	20–33
	(24)	(23)	(24)	(25)	(24)	(24)
Salinity ($\times 10^{-3}$)						
Baseline data (2000)	39–41	39–41	40–43	40–43.5	39–41	39.5–41.5
	(40)	(40)	(42)	(41)	(40)	(40.5)
Oil spill monitoring (2001)[1]	39–40	39–41	39–40	39–41	40–41	39–41
	(40)	(40)	(39)	(40)	(40)	(40)
Post-spill monitoring (2002)	39–40	39–41	41–44	41–43	39–41	39–41
	(40)	(39.5)	(42)	(41)	(39.5)	(40)
pH						
Baseline data (2000)	8.0–8.1	8.1–8.2	7.9–8.2	8.0–8.1	8.1–8.2	8.1–8.2
	(8.0)	(8.1)	(8.1)	(8.0)	(8.0)	(8.1)
Oil spill monitoring (2001)[1]	8.2–8.3	8.1–8.4	8.2–8.3	8.1–8.2	8.1–8.2	8.0–8.1
	(8.2)	(8.2)	(8.2)	(8.1)	(8.1)	(8.0)
Post-spill monitoring (2002)	8.0–8.1	7.9–8.1	7.9–8.1	7.9–8.0	8.1–8.2	8.0–8.2
	(8.0)	(8.0)	(8.0)	(8.0)	(8.1)	(8.1)

(continued)

Table 1. (continued)

Parameters	Off Mamzar	Off Burj Al Arab	Off DEWA	Off DUBAL	Off Jabal Ali	Off Ras Hisyan
NO_3-N ($\mu g\,l^{-1}$)						
Baseline data (2000)	40–70	70	40	40	40	40–70
	(55)	(70)	(40)	(40)	(40)	(55)
Oil spill monitoring (2001)[1]	50	120	180	50	70	40
	(50)	(120)	(180)	(50)	(70)	(40)
Post-spill monitoring (2002)	40–40	40–400	40–530	40–220	40–40	40–40
	(40)	(144)	(163)	(95)	(40)	(40)
PO_4-P($\mu g\,l^{-1}$)						
Baseline data (2000)	–	–	–	–	–	10
	–	–	–	–	–	(10)
Oil spill monitoring(2001)[1]	20	20	20	150	20	20
	(20)	(20)	(20)	(150)	(20)	(20)
Post-spill monitoring (2002)	30–30	20–20	20–50	20–20	20–20	20
	(30)	(20)	(30)	(20)	(20)	(20)
Petroleum Hydrocarbons ($\mu g\,l^{-1}$) as crysene						
Baseline data (2000)	0.3–0.55	0.3–0.3	0.3–0.5	0.3–0.3	0.3–2.6	0.3–0.79
	(0.42)	(0.3)	(0.4)	(0.3)	(1.06)	(0.46)
Oil spill monitoring (2001)[1] (as Total Petroleum Hydrocarbons)	11600	950	500	880	500	500
	(11600)	(950)	(880)	(500)	(500)	(500)
Post-spill monitoring (2002)	0.5–1.1	1.3–3.5	0.5–1.1	5.2	2.9	1.0
	(0.9)	(2.4)	(0.8)	(5.2)	(2.9)	(1.0)

[1] Monitoring of oil spill conducted during 16th and 17th April 2001 from the coastal belt stations 1–6.

Table 2. Biological characteristics along the coastal environment of Dubai during oil spill compared with baseline and post-spill monitoring (baseline and post-monitoring values are given as average).

Parameters	Off Mamzar	Off Burj Al Arab	Off DEWA	Off DUBAL	Off Jabal Ali	Off Ras Hisyan
Chlorophyll-a (mg m^{-3})						
Baseline data (2000)	2.2	0.9	1.2	1.2	1.2	1.4
Oil spill monitoring (2001)	0.5	0.5	1.0	1.6	1.2	2.4
Post-spill monitoring (2002)	2.2	1.0	0.9	1.0	–	2.5
Phytoplankton (cell counts no. $\times 10^3)^{-1}$						
Baseline data (2000)	28.2	30.0	44.1	24.7	25.0	19.7
Oil spill monitoring (2001)	10.6	11.6	32.8	25.2	24.2	20.6
Post-spill monitoring (2002)	38.6	22.2	28.4	24.6	18.4	28.6
Total species (no.)						
Baseline data (2000)	20	20	22	25	26	26
Oil spill monitoring (2001)	12	12	14	18	20	20
Post-spill monitoring (2002)	22	20	20	20	22	20
Major Species						
Baseline data (2000)	*Rhizosolenia*, *Leptocylindrus* *Nitzschia*, and *Ceratium*	*Rhizosolenia*, *Leptocylindrus* *Nitzschia*, and *Prorocentrum*	*Rhizosolenia*, *Chaetoceros* *Nitzschia*, and *Ceratium*	*Rhizosolenia*, *Chaetoceros* *Nitzschia*, and *Ceratium*	*Nitzschia*, *Chaetoceros* *Rhizosolenia*, and *Ceratium*	*Rhizosolenia*, *Nitzschia* *Coscinodiscus*, and *Ceratium*
Oil spill monitoring (2001)	*Rhizosolenia*, *Nitzschia* *Ceratium*, and *Leptocylindrus*	*Rhizosolenia*, *Nitzschia Ceratium*, and *Coscinodiscus*	*Rhizosolenia*, *Chaetoceros* *Nitzschia*, and *Ceratium*	*Rhizosolenia*, *Nitzschia* *Chaetoceros*, and *Ceratium*	*Rhizosolenia*, *Leptocylindrus* *Nitzschia*, and *Chaetoceros*	*Rhizosolenia*, *Leptocylindrus* *Nitzschia*, and *Chaetoceros*
Post-spill monitoring (2002)	*Nitzschia*, *Rhizosolenia*, *Leptocylindrus*, and *Ceratium*	*Rhizosolenia*, *Leptocylindrus* *Nitzschia*, and *Ceratium*	*Rhizosolenia*, *Leptocylindrus* *Chaetoceros*, and *Nitzschia*	*Rhizosolenia*, *Chaetoceros* *Nitzschia*, and *Ceratium*	*Rhizosolenia*, *Leptocylindrus* *Nitzschia*, and *Ceratium*	*Rhizosolenia*, *Chaetoceros* *Leptocylindrus*, and *Ceratium*

(continued)

Table 2. (continued)

Parameters	Off Mamzar	Off Burj Al Arab	Off DEWA	Off DUBAL	Off Jabal Ali	Off Ras Hisyan
Zooplankton biomass [ml (100 m^3)]$^{-1}$						
Baseline data (2000)	2.2	1.6	2.2	2.7	1.5	1.9
Oil spill monitoring (2001)	2.2	1.5	2.2	2.0	2.2	2.5
Post-spill monitoring (2002)	3.0	1.8	2.0	1.7	2.2	2.5
Zooplankton population [no. × 10^3 (100 m^3)						
Baseline data (2000)	40	56	58	48	80	86
Oil spill monitoring (2001)	38	34	52	49	64	40
Post-spill monitoring (2002)	80	36	46	22	24	22
Zooplankton total groups (no.)						
Baseline data (2000)	10	9	10	12	10	9
Oil spill monitoring (2001)	8	6	10	11	11	11
Post-spill monitoring (2002)	11	10	12	9	10	10
Major Groups						
Baseline data (2000)	Copepods, Decapods, and Chatognthas	Copepods, Cladocerans, and Gastropods	Copepods, Chaetognths, and Gastropods	Copepods, Lucifer's, and Decapods	Copepods, Chaetognths, and Lucifer's	Copepods, Chaetognths, and Lucifer's
Oil spill monitoring (2001)	Copepods, fish egg, and Decapods	Copepods, Decapods, and fish larvae	Copepods, fish egg, and fish larvae	Copepods, Chaetognaths, and Gastropods	Copepods, fish egg, and fish larvae	Copepods, fish egg, and Lucifer's
Post-spill monitoring (2002)	Copepods, Decapods, and fish larvae	Copepods, Chaetognths, and Lucifer's	Copepods, Decapods, and fish eggs	Copepods, Decapods, and fish eggs	Copepods, Decapods, and fish larvae	Copepods, Decapods, and fish larvae

Chlorophyll-a varied from 0.9 to 2.2 mg m^{-3} during 16th and 17th April 2001 and this fell within the range of the baseline (1.0–3.2 mg m^{-3}) and post-monitoring results (0.9–2.5 mg m^{-3}). The slightly low value obtained from the stations 1 and 2 were due to the high impact of oil on the phytoplankton which shrink the phytoplankton species. Petroleum spill can cause immediate adverse impact on phytoplankton by retarding photosynthesis. The levels of post-spill monitoring were well comparable with the normal values.

Phytoplankton species from different class were observed in the samples collected during the present study. The cell counts were comparable with the baseline at stations 3–6 whereas stations 1 and 2, recorded relatively very low phytoplankton standing stock and species diversity. In addition to this, many species observed at these locations were shrunken and coated with oil (Fig. 2). *Rhizosolenia stiliformis*, *R. imbricata*, *R. stolterfothi*, *Nitzschia*, *Prorocentrum micans*, *Coscinodiscus*, and *Leptocylindrus danicus* were damaged with a coating of black layer of oil. Maximum damaged of the phytoplankton in the samples collected from stations 1 and 2 were seen particularly in the dinoflagellates species followed by the diatoms. These stations also show very high concentration of TPH, which leads to the chronic conditions in the pelagic organisms. As per available reports, damage to plants can occur during acute phases of spills. Such effects have been reported for phytoplankton and attached algae (Mironov, 1970). Also phytoplanktons are sensitive to oil spill and show different impacts in major groups and sometimes within the species. Green algae are more sensitive to the hydrocarbons than blue green and flagellates (Mironov, 1970).

The impacts of phytoplankton on the bottom were absolutely insignificant. This is usually reported in many of the world oil spills where the impacts of plankton below 15–20 m down were insignificant (Cabioch et al., 1978; Conan, 1982). The post-spill monitoring observed rejuvenate status of the phytoplankton standing stock and species diversity.

The observed zooplankton biomass, total population, groups diversity, and faunal composition between stations 3 and 6 agreed well with the baseline. The highest biomass recorded at station 6 during oil spill was due to predominance of fish eggs and fish larvae. The highest population was obtained at station 5. The major impacts on zooplankton were also encountered at stations 1 and 2. Copepods, *Lucifer* sp., fish eggs, decapods larvae, fish larvae, and chaetognaths were the major groups of zooplankton seen affected at these locations.

Microscopic examination of the zooplankton at stations 1 and 2 revealed black residue of hydrocarbons in gut content of copepods (Fig. 2). Laboratory studies on plankton species demonstrate that water-soluble extracts of oils can kill planktonic organisms (Davenport, 1982). Parker (1970) also reported hydrocarbons in the gut contents of copepods. Patches of hydrocarbons are also seen on the *Lucifer* sp., fish eggs and fish larvae in the samples of zooplankton collected from stations 1 to 3. Floating eggs and larvae close to the sea surface are likely to be impacted by the droplet and hydrocarbons that leached into water column from the oil slick. Some modeling studies (Reed et al., 1984; Hulburt et al., 1991) have indicated that fish larvae have distinct potential to affect recruitment in fish populations.

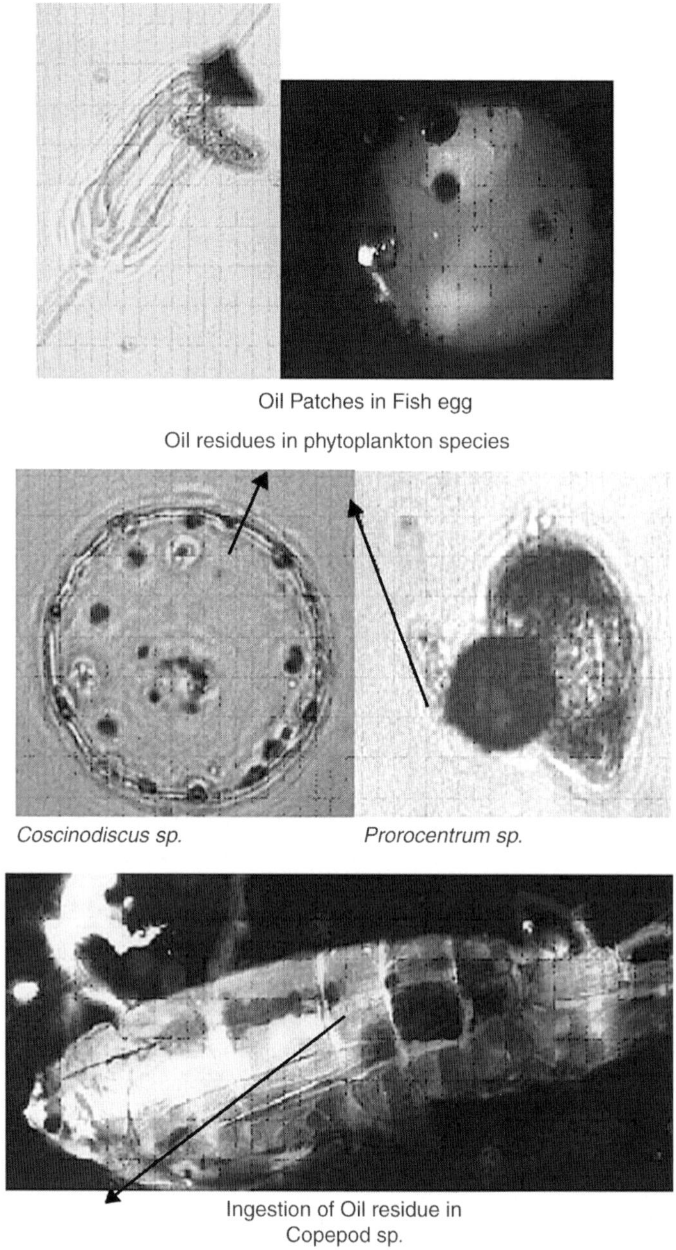

Figure 2. Oil patches in some species of phytoplankton and zooplankton.

4. Conclusion

The present investigation revealed a clear impact of oil spill on local environmental conditions, especially, en route of its spread along with water movement. Levels of nutrient, pH, salinity, and temperature of water were all within normal ranges with occasional increase due to local environmental conditions. The effect was more significant on the level of hydrocarbon content in water though the effect of sediment biota was not evident. The pronounced effect of spill from *Zainab* was observed in the phytoplankton standing stock and diversity, which shows damaged species. Effect on secondary trophic level shows ingestion of petroleum hydrocarbons in the zooplankton samples particularly in copepods and oil droplets on fish eggs and larvae. Post-monitoring indicates relatively fast recovery, which restored the area at previous level within 1 year.

References

Al-Abdali, F., Massoud, M.S., Al Ghadban, A.N., 1986. Bottom sediment of the Arabian Gulf III. Trace metal contents as indicator of pollution and implications for the effects and fate of Kuwait Oil Slick. Environ. Pollut. 93, 285–301.

Al-Ghadban, A.N., Massoud, M.S., Abdali, F., 1996. Bottom sediments of the Arabian Gulf I. Sedimentological characteristics. Kuwait J. Sci. Eng. 23, 71–78.

American Public Health Association (APHA), 1995. American Water Works Association, Water Pollution Control Federation 1971, 19th edn. Standard Methods for Examination of Water and Wastewater. American Public Health Association, New York, NY.

Cabioch, L., Dauvin, J.C., Gentil, F., 1978. Preliminary observations on pollution of the seabed and disturbance of sub littoral communities in northern Brittany by oil from Amoco Cadiz. Mar. Pollut. Bull. 9, 303–307.

Capuzzo, J.M., 1987. In: Bosesch, D., Rabalais, N. (Eds), Long-term Environmental Effects of Offshore Oil and Gas Development. Elsevier Applied Science, London, p. 343.

Conan, G., 1982. The long-term effect of the Amoco Cadiz oil spill. Phil. Trans. R. Soc. Lond. B 297, 323–333.

Davenport, J., 1982. Phil. Trans. R. Soc. Lond. B 297, 369.

EI-Samara, M., Emara, H.I., Shunbo, F., 1986. Dissolved petroleum hydrocarbon in the northwest Arabian Gulf. Mar. Pollut. Bull. 19, 14–18.

Fowler, S.W., 1985. Coastal baseline studies of pollutants in Bahrain, UAE and Oman. Proceedings of a Regional Symposium for the Evaluation of Marine Pollution Monitoring and Research Programmes, December 8–11, Al Ain, UAE ROPME/GC-4/2, pp. 155–180.

Hinrichsen, D., 1996. Living on Edges: Coasts in Crises. Island Press, Washington, DC.

Hulburt, S.E., French, D., Taylor, B.J., 1991. Evaluation of the Potential Effects of Major Oil Spills on Grand Band Commercial Fish Species as a Result of Impacts on Eggs and Larvae. Report prepared for Environmental Studies Research Fund by SAS Consulting Ltd. Box 2625, Dartmouth East, Nova Scotia B2W 3X8 (Canada).

KFUPM/RI, 1987. Marine studies in Saudi Arabian waters for the Arabian Gulf. Research Institute, King Fahad University of Petroleum and Minerals, Dhahran, Saudi Arabia.

Literathy, P., Ali, M., Al-Harmi, L., Joseph, P., Lee, J., 1986. Monitoring of Oceanographic Characteristics and Baseline Studies of Oil and Non-Oil Pollutants in the Territorial Waters of Kuwait. Project Report, Kuwait Institute of Scientific Research, Kuwait.

Mahmoudi, S., 1997. Legal protection of the Persian Gulf's marine environment. Mar. Policy 21, 53–62.

Massoud, M.S., Al-Abdali, F., Al-Ghadban, A.N., 1998. The status of oil pollution in the Arabian Gulf by the end of 1993. Environ. Int. 24, 11–22.

Mironov, O.G., 1970. In: Procedings of the FAO conference on marine pollution and its effect on living resources and fish, (FAO Rome) F-92.

MOOPAM, 1999. Manual of Oceanographic Observation and Pollutant Analysis Methods. Regional Organization of Protection for the Marine Environment, Kuwait.

Otsuki, A., Abdulraheem, M.Y., Reynold, R.M., 1998. Offshore Environment of the ROPME Sea Area After War-Related Oil Spill: Results of the 1993–94 Umitak-Maru Cruises. Terrs Scientific Publishing Company, Tokyo.

Parker, C.A., 1970. AMI Report B 198 CMS.

Reed, M., Spaulding, M.L., Lorda, E., Walker, H., Saila, S.B., 1984. Oil spill fishery impact assessment modeling the fisheries recruitment problem. Estuar. Coast. Shelf Sci. 19, 591–610.

Reynold, M., 1993. Physical oceanography of the Gulf, Strait of Hormuz and Gulf of Oman: results from the Mt Michell expedition. Mar. Pollut. Bull. 27, 35–60.

ROPME, 1999. Regional Organization for the Protection of the Marine Environment. Regional Report of the State of Marine Environment ROPME, Kuwait.

Strickland, J.D.H., Parsons, T.R., 1972. A practical handbook of seawater analysis. J. Fish. Res. Board Can. Bull. 167, 311.

V

Combating, prevention and treatment of oil pollution

Oil Pollution and its Environmental Impact in the Arabian Gulf Region
M. Al-Azab, W. El-Shorbagy and S. Al-Ghais, editors

Chapter 14

Oil pollution in the ROPME sea area – prevention, abatement, combating

Hassan Mohammadi

ROPME, P.O. Box 26388, Safat 13124, State of Kuwait, Kuwait

Abstract

The ROPME Sea Area (RSA) is considered to be an area with one of the highest oil pollution risks in the world. This is mainly due to the concentration of offshore installations, tanker loading terminals, and the huge volume and density of marine transportation of oil.

The operational and accidental oil pollution is a major challenge in the RSA. The impacts of offshore installations of oil and gas particularly those of produced water on the marine environment in general and in shallow waters or near to ecologically sensitive areas in particular are noticeable. The operational pollution from ships and dumping of ballast water are also among the main causes of chronic oil pollution in the Region. At the same time, there are a great number of marine emergency cases resulting in substantial pollution or imminent threat of substantial pollution to the marine environment by oil from collisions, strandings and other incidents involving oil tankers, blow-outs arising from petroleum drilling and production activities, and the presence of oil from the failure of industrial installations. Such marine emergencies as published in the Oil Spill Intelligence Report indicate that 30% of oil spills greater than ten million gallons worldwide are related to the ROPME Region. Another major concern of environmental authorities is the introduction of invasive species that could threaten ecosystems, habitats, and species. To this effect, the establishment of reception facilities for oily wastes and other wastes, and the application of all appropriate measures for the control and management of ships' ballast water and sediments are now on the high priority agenda to be pursued attentively.

To address the environmental challenge of oil pollution, a series of program activities for the prevention, abatement, and combating of land-based and sea-based sources of marine oil pollution have been developed and implemented in accordance with the provisions of the Kuwait Regional Convention and its Protocols. Accordingly, marine oil emergency preparedness and response are dealt with by the Emergency Protocol, offshore oil pollution prevention abatement, and control are addressed by the Continental Shelf Protocol, land-based sources of oil pollution are covered by the LBA Protocol, and transportation and disposal of wastes such as ballast water and oily wastes from ships and the establishment of reception facilities are addressed by the Transboundary Protocol. Other areas of focus have been the application of EIA for major development projects, the use of satellite receiving and processing system for monitoring of marine and coastal areas, the oil emergency preparedness and response, and a comprehensive program for training and capacity-building. All these programs are the central objectives of the Kuwait Action Plan (KAP), and have been carefully pursued by way of a coordinated National and Regional process-oriented framework of action.

This chapter also describes the institutional arrangements developed for the implementation of the regional instruments. To this effect, liaison has been established and maintained between ROPME and the Contracting States, as well as with the competent regional and international organizations.

E-mail address: ropme@qualitynet.net (H. Mohammadi).

Such arrangements have been made to ensure that cooperation and coordination of action are achieved on a regional basis.

1. Introduction

The growing concern about the danger of marine pollution by oil in early 1950's led to the adoption of the International Convention for the Prevention of Pollution of the Sea by Oil in 1954. However, this Convention was primarily aimed at pollution resulting from routine tanker operations and oily wastes from machinery spaces and did not deal with tanker accidents or other sources of marine pollution.

The Torrey Canyon incident off south-west coast of England in March 1967 resulted in 117,000 metric tons of oil spill causing pollution damage of an extent hitherto unknown. This incident made the world aware of the need for contingency planning and the international regimes for the purpose of responding to marine emergencies. To this effect, the Civil Liability Convention (1969), Intervention Convention (1969), Fund Convention (1971), MARPOL 73/78, and OPRC (1990) were adopted to protect against and compensate for similar accidents.

Although the international conventions have made a substantial positive drive towards decreasing the amount of oil that enters the sea; however, oil pollution has been and still is a continuous phenomenon in the marine environment. This has been particularly noted in the ROPME Region urging the adoption of a number of legal instruments in compliance with and in support of the provisions of the international conventions. Needless to say, these regional instruments are described in this chapter have made the mandate more specific and have had an important role in harmonizing the policies and the programs of Contracting States concerning protection of the marine environment from oil pollution under the national jurisdiction of each State and that of the Region.

2. Regional mechanisms for oil pollution control

The coastal States of the Region with the assistance of UNEP developed the following three legal documents, which were adopted at the Regional Conference of Plenipotentiaries on the Protection and Development of the Marine Environment and the Coastal Areas, held in Kuwait from 15 to 23 April 1978:

(i) Kuwait Regional Convention for Cooperation on the Protection of the Marine Environment from Pollution, 1978.
(ii) Protocol concerning Regional Cooperation in Combating Pollution by Oil and other Harmful Substances in Cases of Emergency, 1978.
(iii) Kuwait Action Plan (KAP) for the Protection and Development of the Marine Environment and the Coastal Areas, 1978.

The preparation of additional Protocols as recommended under the Legal Component of the KAP has also been pursued with the adoption of the following Protocols:

(i) Protocol concerning Marine Pollution resulting from Exploration and Exploitation of the Continental Shelf, 1989.

(ii) Protocol for the Protection of the Marine Environment against Pollution from Land-Based Sources, 1990.
(iii) Protocol on the Control of Marine Transboundary Movements and Disposal of Hazardous Wastes and Other Wastes, 1998.

3. Kuwait Action Plan

The KAP sets forth a framework for an environmentally sound and comprehensive approach to regional development. The assessment of the origin and magnitude of oil pollution in the Region is one of the central objectives of KAP, comprising:

(a) baseline studies on the sources, transport, and distribution of oil and petroleum hydrocarbon pollution in the Region;
(b) physical, chemical, and biological oceanography of the Region relevant to the transport, distribution and fate of oil as a pollutant;
(c) marine meteorology relevant to the transport and distribution of oil as a pollutant.

The following cooperative programs are also fully addressed under the Environmental Management Component of KAP:

(a) formulation of regional contingency plans for accidents involving oil exploration, exploitation and transport, and strengthening the meteorological services contributing to the development of contingency plans and to their execution in coordination with existing or future marine regional meteorological programs;
(b) upkeep of records of oil pollution incidents in the Region with relevant information on the impact of such pollution on the marine environment.

4. Protocol concerning Regional Cooperation in Combating Pollution by Oil and other Harmful Substances in Cases of Emergency, 1978

The ROPME Sea Area (RSA) is considered to be an area with one of the highest oil pollution risks in the world. This is mainly due to the concentration of offshore installations, tanker-loading terminals, and the huge volume and density of marine transportation of oil.

According to the Oil Spill Intelligence Report, out of 20 cases of oil spills greater than ten million gallons (@ 34,000 metric tons) worldwide, six cases have been related to the ROPME Region. The 1981 spill from storage tanks in Shuaiba (31,170,000 gallons or @ 106,000 metric tons), the 1985 spill of tanker Nova, 140 km south of Kharg Island (21,350,000 gallons or @ 73,000 metric tons), the 1983 spill from tanker Assimi at Ras Al-Hadd, 58 miles from Muscat (15,800,000 gallons or @ 54,000 metric tons), and the 1983 tanker Pericles GC, 30 km north-east of Doha (14,000,000 gallons or @ 48,000 metric tons) are among the major spills recorded for the Region.

Smaller scale oil pollution incidents are also numerous in the RSA. The Ras Tannura submarine pipeline puncture (1979), the Hasbah 6 well blow-out (1980), the Texaco Caribbean tanker at Fujairah (1987), the tanker Akari fire at Jebel Ali (1987), the motor tanker Enterprise at Mina Al-Ahmadi (1989), the explosion of tanker Surf City in Abu Musa Island (1990), the spill of tanker Koyali at UAE (1990), and the collision of tanker

Seki with the UAE vessel BAYNUNA at Fujairah (1994) are just a few examples to highlight the fact that the threat of oil spill in the Region is a continuous event.

Environmental catastrophes may also be caused by deliberate destruction of refineries, loading terminals, oil storage tanks, pipelines, sinking of oil tankers, as well as blow-out of oil wells for hostile purposes. In this connection, the Sea Area has suffered from two unprecedented, devastating oil pollution incidents within a decade; first during the Iran–Iraq war and then through the War of 1991. Massive oil spill caused by intentional destruction of oil wells and related installations has been a recent development and happened to be the first of its kind in the annals of international affairs. The Nowruz incident is estimated to have released an amount of 1,870,000 barrels or 292,000 metric tons of oil from February to December 1983. However, due to the burning and high water temperatures, more than 50% of this amount evaporated and remaining oil of approximately 150,000 metric tons spilled into the Sea Area. In the second incident, the aggression on the environment of Kuwait resulted in the largest oil spill in history (6–8 million barrels), the long-term detrimental effects of which are still to be investigated.

4.1. Protocol provisions

The objective of the Protocol, which was signed on 24 April 1978 and entered into force on 1 July 1979, is to provide cooperative and effective preventive and response measures to deal with marine emergencies caused by oil and other harmful substances. Marine emergency means any casualty, incident, occurrence or situation, however caused, resulting in substantial pollution or imminent threat of substantial pollution to the marine environment by oil or other harmful substances and includes, *inter alia*, collisions, strandings and other incidents involving ships, including tankers, blow-outs arising from petroleum drilling and production activities, and the presence of oil or other harmful substances arising from the failure of industrial installations.

4.2. General obligations

Each Contracting State is to direct its appropriate officials to require masters of ships, pilots of aircraft, and persons in charge of offshore platforms and other similar structures operating in the marine environment and under its jurisdiction to report the existence of any marine emergency in the Sea Area to the appropriate national authority and to the Center. Any Contracting State receiving such a report shall promptly inform the following of the marine emergency:

- The Center;
- All other Contracting States;
- The flag State of any foreign ship involved in the marine emergency concerned.

Contracting States are also required to cooperate and coordinate their activities, *inter alia*, on the following:

- Distribution and allocation of stocks of material and equipment;
- Training of personnel for marine emergency response;

- Marine Pollution surveillance and monitoring activities;
- Methods of communication in respect of marine emergencies;
- Facilitation of the transfer of personnel, equipment and materials involved in marine emergency responses into, out of, and through the territories of the Parties;
- Other matters to which the Protocol applies.

5. Marine Emergency Mutual aid Center

In conformity with the provisions of the Protocol, the Marine Emergency Mutual Aid Center (MEMAC) was established in Bahrain and started functioning in March 1983. The main objective of MEMAC/Center is to strengthen the capacities of the Contracting States and to facilitate cooperation among them in order to combat pollution by oil and other harmful substances in cases of marine emergencies. Another objective is to assist Contracting States in the development of their own national capabilities to combat pollution by oil and other harmful substances and to coordinate and facilitate information exchange, technological cooperation and training. The other objective is to initiate operations to combat pollution by oil and other harmful substances at the regional level.

MEMAC is an intergovernmental emergency center with the mandate to coordinate the regional planning, programming, monitoring, exchange of information and data, training, capacity-building, as well as developing regional arrangements for transboundary movements of personnel, equipment and materials in cases of emergency.

In practice, local pollution is dealt with the oil company and is coordinated by the Regional Clean Sea Organization (RECSO); pollution of a wider scope and those which do not fall within the jurisdiction of oil companies are handled by the national authorities through the National Contingency Plans; and massive oil spills, the size and scope of which are beyond the capabilities of individual States or oil companies, fall within the purview of the Marine Emergency Protocol and are coordinated by MEMAC.

RECSO is an oil industry organization for coordinating the emergency responses of member companies. In fact, RECSO members collectively possess a large number of oil spill combating equipment, as well as technical capabilities, and resources which could be made available to other oil spill emergencies in the Region.

5.1. National contingency plan

In accordance with Article II of the Protocol, Contracting States are required to maintain and promote their contingency plans and means for marine emergency response. The means shall include, in particular, the available anti-pollution equipment and materials, ships, aircraft, and manpower for operations in oil spill emergencies. Each National Contingency Plan is to clearly specify the coordination mechanism and responsibilities of various national entities, such as oil companies, port authorities, coast guards, navy, shipping authorities, environment protection bodies, as well as other concerned institutions whose technical and manpower support is essential for responding to any major marine oil pollution incident. The 1991 oil spill catastrophe made it clear that certain Member States did not have even the minimum response capabilities.

5.2. Regional contingency plan

The capacity of the Region to respond to marine emergencies is not keeping pace with the developments and contingencies of oil pollution challenges. Among the major obstacles for an effective marine emergency response in the Region, is lack of an established regional coordination mechanism for effective cooperation among Member States in times of emergencies, shortage of technical expertise, inadequacy of response equipment and absence of a proper communication/information system for provision of the required information on emergency response operations.

There is a need to mobilize all the resources, strengthen the response capabilities and conclude a comprehensive Regional Contingency Plan to facilitate cooperation in preparing for and responding to major oil pollution incidents in the Region. Elements of the National Contingency Plans and the contingency plans developed by RECSO member companies are to be incorporated into the Regional Contingency Plan. The full-fledged cooperation between MEMAC and Member States and between RECSO and MEMAC will serve the purpose.

MEMAC has exerted all possible efforts for the protection of the Sea Area since establishment despite enormous hindrances in the path of the Center. However, the tasks are massive and the extent of prevailing pollution and environmental degradation by oil are far reaching and beyond the existing capabilities of the Region. Large areas of our shores are heavily polluted by oil and sensitive marine ecosystems are at danger. There is an urgent need to join hands and strengthen the emergency response capabilities of the Region in order to implement the comprehensive Regional Contingency Plan as prepared by MEMAC.

5.3. GIS and remote sensing

Remote sensing and the application of satellite-based technologies for monitoring the environmental phenomena, and for pursing claims due to offences on marine environment specially those of oil spills from tankers and discharges from land-based activities and offshore operations is of growing interest. To this effect, ROPME has continuously upgraded the remote sensing capabilities and has recently established a small Satellite Receiving Station. The Station has been operational ever since the establishment in January 2003 enabling ROPME to provide structured remote sensing programs/services on national priorities to Member States. We now need to develop regional networking for the exchange of data, information, and test sites measurements related to specific national requirements. Phase 2 of the Receiving Station for feasible options towards upgrading the system is also under consideration.

5.4. Codes of practice for the use of oil spill dispersants in the ROPME sea area

The code of practice is based on the IMO/UNEP Guidelines on Oil Spill Dispersant Application including Environmental Considerations (IMO 1995) and the International Petroleum Industry Environmental Conservation Association (IPIECA) documents, and is amended for the operational and environmental considerations specific for use of oil spill chemicals in the RSA. The code of practice provides all those concerned with oil spill response in the Region with useful information and advice on oil spill response options, contingency planning and the use of dispersants. The following conclusions are also made in the code of practice:

(i) Mechanical oil containment and collection equipment should represent the first line of defense applied in oil spill incidents. The use of oil dispersants and other oil spill chemicals is to be considered when mechanical containment fails and the environmental conditions permit such a use.
(ii) Until such time as a specified approval procedure including toxicity and effectiveness testing specific to the marine organisms and conditions of the RSA is developed, the approved status of oil spill products will continue to be given on the basis of the approval of the internationally recognized institutions, i.e. CEDRE-France, DEFRA-UK and US-EPA.
(iii) Only concentrate dispersants will be approved. To be placed on the ROPME List of Approved Oil Spill Chemicals, dispersants must be on the current list of at least two of the three institutions. Certified toxicity and effectiveness test results, as well as information on the physical and chemical properties of dispersants must be submitted with the application.
(iv) Shoreline cleaning agents which may come under different classification or categories (e.g. surface washing agents, shoreline cleaners, surface cleaners or even as dispersants) must be approved by two of the three institutions.
(v) Due to the limited information available regarding the effectiveness and fate of biological additives, as well as the potential risks of adversely affecting the natural biodiversity when living organisms are introduced, it is recommended that such products, at present, not to be introduced in the Region.
(vi) As the effect of absorbents on spilled oil is physical, rather than chemical, it is considered that approval is not needed for these products. There is therefore no need to include them on the approved list.
(vii) Emulsion breakers, surface film chemicals, surface collecting agents, and miscellaneous oil control agents are excluded from the ROPME List because they are no longer available or the technology has been superseded by more modern developments.
(viii) ROPME is to update the List of Approved Oil Spill Chemicals continuously and to revise the List once every 3 years, taking into consideration the international developments and regional experience. Dispersants that do not meet the requirements will be omitted from the List, although stockpiles of dispersants that were previously approved will still be permitted to be used until the shelf life.

5.5. MEMAC programme activities

The list of programme activities of MEMAC for the years 2003–2004 is as follows:

(i) Regional Contingency Plan
   - Training Workshop on Contingency Planning
   - Oil Spill Emergency Drills and Oil Spill Response at Managerial, Supervisor and Operator Levels

- Oil Pollution Manual
- NOAA Trajectory Model and other Models

(ii) Guidelines for Pollution Damage Assessment and Manual for Compensation Claims
(iii) Emergency Fund for Protection of Marine Environment
- Proposal for establishing an Oil Response and Salvage Corporation

(iv) Development of Port State Control Procedures
(v) Safety Programme for Harmful/Hazardous Substances
(vi) Annex to the Emergency Protocol on Combating Pollution
(vii) Periodical Reports on Marine Emergency
(viii) Establishing Reception Facilities in the Region
(ix) Convention on Oil Pollution Preparedness, Response and Cooperation, 1990 (OPRC)
- Establishing Marine Search and Rescue Centers.

(x) Establishment of Data-Base
(xi) Survey of Wrecks
(xii) Marine Environment High-Risk Areas.

6. Protocol concerning Marine Pollution Resulting from Exploration and Exploitation of the Continental Shelf, 1989

The Protocol was concluded in 1988, signed on 29 March 1989, and entered into force on 17 February 1990. Later, the following Guidelines to the Protocol were adopted by the Seventh Meeting of ROPME Council on 21 February 1990:

(i) Guidelines on requirements for environmental impact surveys and assessments.
(ii) Guidelines on the use and storage of chemicals in offshore operations.
(iii) Guidelines on the conduct of seismic operations.
(iv) Guidelines on disposal of drill cuttings on the sea-bed.

These Guidelines are to assist Contracting States in developing their specific plans and measures in compliance with the provisions of the Protocol. Also, the application of common standards, criteria and regulations, as well as the harmonization of environmental policies, programs, administration and legislation of Contracting States for the fulfillment of their obligations under the Protocol, are major objectives to be achieved in the near future.

6.1. Protocol provisions

The objective of the Protocol is to coordinate regional activities towards protection of the marine environment against pollution from exploration and exploitation of oil and gas in the continental shelf. The Protocol with fifteen articles and four Guidelines is a broad framework for developing comprehensive action plans delineating the obligations of Contracting States at the national and regional levels for sound environmental practices in offshore exploration and production (E&P) activities.

The offshore installations of oil and gas are located in the inner RSA, which is badly suffering from extremes of salinity, temperature, and oil pollution. The evaporation is high, the precipitation is poor and the river discharge is decreasing. The water lost by evaporation of the sea is mainly compensated by water exchange through the Strait of Hormuz. However, the rate of water exchange is low and the retention of pollutants is prolonged. Such being the case, the impacts of offshore operations especially those of produced water on the marine environment in general, and in shallow waters or near to ecologically sensitive areas in particular, are more noticeable. The magnitude of environmental impacts is also subject to the volume and the composition of discharge (i.e. oil, production chemicals, NORM, solids, inorganic salts, and metal salts).

The principal objective of produced water management is to minimize its detrimental impacts on the marine environment. This objective can be achieved by addressing all components of produced water discharges in a comprehensive way. To this effect, the operator is to select the least hazardous chemicals in offshore operations. The volume of produced water and the concentration of pollutants should be reduced, and the volume of discharges to be minimized. Re-injection option should be investigated and applied wherever feasible. The re-injected water should be free from additives and production chemicals.

ROPME is advocating the use of the best available and economically feasible technology for the prevention, abatement, and control of marine pollution from offshore operations. Here, the obligation for the best available technology is not absolute as it simply requires the operators to take into account that technology which is in the meantime subject to economic feasibility.

6.2. Licensing of offshore operations

The term "license" is defined in the Protocol for any form of license, permit, or authorization, formally issued under the authority of a Contracting State for undertaking an offshore operation.

Each Contracting State shall ensure that every offshore operation is conducted under a license, which may be granted subject to such conditions for the protection of the marine environment and coastal areas as the Competent State Authority sees fit to impose. The use of license conditions is easily justified since the requirements for control will vary with the change of circumstances. Also, the Competent State Authority shall require the operator to comply with relevant laws and regulations issued under the authority of the State.

6.3. Environmental impact assessment

Environmental impact assessments are considered as necessary to ensure that informed decisions are made and appropriate measures are taken to prevent or mitigate substantial pollution from offshore operations to the Protocol Area.

Each Contracting State shall take necessary measures to ensure that before licensing any offshore operation which could cause significant risks of pollution in the Protocol Area or any adjacent coastal area, an assessment of the potential environmental effects is made. No such operation shall commence until a statement of those effects is submitted and the Competent State Authority is satisfied that the operation will entail no acceptable risk of damage in the Protocol Area or any adjacent coastal area. Competent State Authorities shall have regard to the Guidelines issued by the Organization in determining the scope of any environmental impact statement.

A summary of the environmental impact statement is to be sent to ROPME Secretariat for consideration and necessary follow-up.

6.4. Citing of offshore installations

The citing of offshore installations in the Protocol Area need to be carefully planned in order to protect other related interests, and the rights and duties of other States.

Contracting States shall ensure that offshore operations shall not cause unjustifiable interference with lawful navigation, fishing or any other legitimate activity, and that in citing an installation, due regard shall be had to existing pipelines and cables. Regard shall also be had to the need for protecting sites of special ecological and cultural interest.

Each operator of an offshore installation is to survey the sea-bed in the vicinity of the installation, and remove any debris resulting from the operations which might interfere with lawful fishing.

6.5. Safety provisions

Safety measures are to be undertaken with regard to design, construction, placement, equipment, marking, operation, and maintenance of offshore installations so that they can withstand the stresses imposed by natural forces in the most severe and adverse conditions. To this effect, Contracting States shall take practicable measures to ensure, inter alia, the following:

- Every offshore installation to be used is certified by a Certifying Authority or its designee that is safe and fit for the purpose for which it is to be used.
- Each operator of an offshore installation shall at all times have available, in good working order, equipment and devices to minimize the risk of accidental pollution, and to facilitate prompt response to a pollution emergency.
- Blow-out preventors and other safety equipment are tested, and exercises in their operation are carried out periodically.
- Offshore installations above sea level carry light and other warning instrument, in accordance with international maritime practice.
- All persons engaged in offshore operations shall have had or be given training in accordance with good oil field practice.

6.6. Contingency planning

Each operator of an offshore operation is to have an approved "contingency plan" to deal with any event which may cause significant pollution to the marine environment, and shown to the satisfaction of the Competent State Authority that sufficient expertise and resources are available to put the plan fully into operation. Such contingency plans are to be coordinated with existing national or local contingency plans, and any plans prepared by the Marine Emergency Mutual Aid Center.

6.7. Operational discharges

Discharges of oily wastes and oily water such as discharges of production water, displacement water, processing drainage, platform drainage, machinery space drainage, and well testing occur in the course of normal operation of offshore installations.

Contracting States shall take all practicable measures to ensure the following:

- In the Special Area, no machinery space drainage shall be discharged unless the oil content does not exceed 15 mg/l whilst undiluted.
- No other discharge, expect one derived from drilling operations, shall have an oil content greater than 40 mg/l as an average in any calendar month, and at any time not exceeding 100 mg/l whilst undiluted.
- Discharge point for oily wastes shall be well below the surface of the sea as appropriate.
- All necessary precautions shall be taken to minimize losses of oil into the sea from oil and gas collected or flared from well testing.
- Oil-based drilling fluids shall not be used in drilling operations except in special circumstances and with the express sanction of the Competent State Authority.
- Oil-based drilling fluids shall not be discharged to any parts of the Protocol Area within the national jurisdiction.
- Water-based drilling muds must not contain persistent systemic toxins which may continue to pose an environmental threat after the drilling fluid discharge.

6.8. Sewage and garbage

Disposal of sewage and garbage into the sea is to meet all the requirements of the Protocol. Each Contracting State shall provide at convenient points on its coastline, reception facilities for general garbage from manned offshore installations operating within the Protocol Area under its jurisdiction.

6.9. Drilling and production chemicals

The use of chemicals is necessary for efficient operation in the oil extraction industry, and as such large quantities of chemicals are used in drilling and in production. Article XI of the Protocol and Guidelines to the Protocol are to regulate the use and storage of chemicals in offshore operations. Accordingly, each Contracting State shall take all appropriate measures to ensure that each operator of an offshore installation shall prepare, and submit for approval by the Competent State Authority, a "chemical use plan". Application for amendments to the plan may be submitted subsequently and approved. Also, the operator shall notify the Competent State Authority for the use of a chemical outside the scope of an approved plan that may escape into the marine environment. However, to prevent the risk of injury to person or excessive damage to property in cases of emergency, the notification need not be given prior to the use of the chemical.

The chemical use plan by definition is a plan drawn up by the operator of an offshore installation which shows:

- the chemicals intended to be used in operations;
- the purpose or purposes for the use of the chemicals;
- the maximum concentration of the chemicals intended to be used within any other substances, and maximum amounts intended to be used in any specified period;
- the area within which the chemical may escape into the marine environment.

The Competent State Authority has the power to prohibit, limit or regulate the use of a chemical or product and to impose conditions on its storage and its use, for the purpose of protecting the marine environment. In exercising that power, the Authority shall have regard to the Guidelines issued by the Organization.

ROPME Guidelines on the Use and Storage of Chemicals in Offshore Operations are a useful guide to assist the operators in their submission of chemical use plans. Accordingly, some chemicals are exempt from notification, which means that the operator need not include them in the chemical use plan. For a number of chemicals, approval is deemed to have been granted, provided the Competent State Authority is notified at least 21 days before the offshore storage or use of the chemicals. And still, the use of some listed chemical constituents should not be approved, save in special circumstances.

The need for a chemical use plan is readily justified as a very practicable instrument used for the regulation, limitation, and prohibition of the use and storage of chemicals. This procedure relieves ROPME Contracting States from strict listing and other categorization of chemicals for use in offshore operations.

The Protocol has left the implementation of chemical use plans to Contracting States, and it would be for the States to ensure compliance with Protocol provisions by measures appropriate to their legal framework and administrative structures. However, there are significant variations between the technical capabilities and the enforcement procedures of Contracting States, and as such the results of their control measures on E&P chemicals are different. Therefore, there is a need to revise and harmonize the existing national procedures on E&P chemicals, and to adopt a common application procedure for the approval of chemical use plans in the entire RSA.

6.10. Seismic operations

Seismic operations in the Protocol Area are to observe the Guidelines issued by the Organization. Protocol Guidelines on the Conduct of Seismic Operations provide that each operator to have an approved "seismic operations plan" before commencing seismic operations. All the requirements for a seismic operations plan such as, notice procedure, equipment and manning of vessels, the safeguard for environmental protection, as well as regards to fishing and other interests are elaborated in the Protocol Guidelines.

6.11. Drill cuttings

Guidelines to the Protocol on the Regulations of Drill Cuttings provide guidance for disposal of all types of drill cuttings, i.e. oil-based, alternative oil-based, and water-based drilling fluids used.

Contracting States shall ensure that no drill cuttings are deposited on the sea-bed in a sensitive area except in special circumstances and in accordance with an approval granted by the Competent State Authority. Also, drill cuttings from drilling during which diesel oil-based drilling fluid or alternative oil-based drilling fluid has been used shall not be deposited on the sea-bed in the Protocol Area without the express approval of the Competent State Authority.

6.12. Other provisions

1. Each Contracting State shall ensure that the operator of an offshore installation:
   (a) in the case of a pipeline:
      - to flush and remove any residual pollutants from the pipeline, and
      - to bury the pipeline, or remove part and bury the remaining parts, so as to eliminate any risk of hindrance to navigation or fishing
   (b) in the case of platforms and other sea-bed apparatus and structures, to remove the installation in whole or in part to ensure safety of navigation and in the interests of fishing.
2. When Contracting States have a common interest in fishing grounds in the Protocol Area, they shall endeavor to adopt a common policy on the removal of installations.
3. Contracting States shall ensure that no offshore installation in use has floated at or near the sea-surface, and no equipment from an offshore installation, shall be deposited on the sea-bed of the continental shelf when it is no longer needed.

7. Protocol for the Protection of the Marine Environment against Pollution from Land-Based Sources, 1990

The sudden and haphazard mushrooming of coastal developments in the past few decades has been staggering in the Region. The impacts of land-based sources of pollution on the coastal waters are significant, particularly those of municipal sewage, and industrial effluents from such industries as petroleum refineries, power, desalination, and petrochemical plants. The power and desalination plants contribute about 48%, petroleum refineries 28%, petrochemicals 19%, and other industries 5% of the total volume of industrial effluents on the coastal area of the Region.

To address Land-Based Sources of Pollution in a comprehensive way, the text of the Protocol with 16 Articles and 3 Annexes was prepared, negotiated and finalized in 1989, signed on 21 February 1990 and entered into force on 2 January 1993.

Annex III to the Protocol provides for regional guidelines, regulations, and permits for the release of wastes. Accordingly, regional regulations for the waste discharge and/or degree of treatment should be specific for each kind of source and, if necessary, may be different between existing and new sources.

Regional regulations along with the programs, measures, and the timetables required for the implementation should be developed on a priority basis, *inter alia*, for the following types of wastes:

- Ballast water, slops, bilges, and other oily water discharges generated by land-based reception facilities and ports through loading and repair operations.
- Brine water and mud discharges from oil and gas drilling and extraction activities from land-based sources.
- Oily and toxic sludges from crude oil and refined products storage facilities.
- Effluents and emissions from petroleum refineries.
- Effluents and emissions from petrochemical and fertilizer plants.
- Emissions from natural gas flaring and desulfurization plants.
- Toxic effluents and emissions from industries such as chlor-alkali, primary aluminium production, pesticides, insecticides, and lead recovery plants.
- Dust emissions from major industrial sources, such as cement, lime, asphalt, and concrete plants.
- Effluents and emissions from power and desalination plants.
- Wastes generated from coastal development activities which may have a significant impact on the Marine Environment.
- Sewage and solid wastes.

ROPME pursues as a principal approach the development of a Regional Program of Action to support and reinforce the National Programs of Action for the implementation of the Protocol. The Regional Program of Action is to meet the environmental needs and enhance the environmental capabilities of the Region and is aimed primarily towards implementation of the Protocol by way of a coordinated National and Regional process-oriented framework of action.

8. Protocol on the Control of Marine Transboundary Movements and Disposal of Hazardous Wastes and Other Wastes, 1998

The main objectives of the Protocol are to protect the marine environment of the Protocol Area from detrimental effects of hazardous wastes and other wastes, to assist Contracting States in environmentally sound management of wastes they generate and to enhance cooperation and coordination of action on a regional basis with the aim of controlling the transboundary movements of hazardous wastes. To this effect, the transboundary movements of wastes, the dumping of wastes at sea, the ballast water of oil tankers, and the wastes of commercial ships are covered by the Protocol. In other words, the Protocol deals with the main provisions of three international conventions, i.e. the Basel Convention, MARPOL 73/78, and the London Convention of 1972 and its 1996 Protocol. It has been the intention not to develop too many Protocols and hence not to create too much bureaucracies and obligations for ROPME Member States which they cannot afford.

The promotion of regional cooperation for the establishment and management of reception facilities for the reception and treatment of ballast water and other wastes from ships and for the development of an effective monitoring and surveillance system to detect and control dumping of wastes at sea, are fully addressed in the Protocol.

The subject of ballast water management has been and still is on the high priority list of international environmental agenda. This is more pressing for our marine environment, which is particularly suffering from chronic oil pollution caused by releases from oil industry, ballast water, and operational discharges of ships. Needless to say, oil in seawater represents a considerable threat for marine life, mariculture, industrial water intakes, and human health. To this effect, the protection of water quality in the vicinity of water intakes for desalination plants and other key industries is of prime concern to countries of the region.

In addition to the physical impact of oil on heat exchangers of power and desalination plants, oil in the soluble form or dispersed in the water column can affect the seawater quality in the vicinity of these plants, as well as the quality of the desalinated water produced. Reaction of hydrocarbons with chlorine, which is injected to protect plants from bio-fouling, produces halocarbons which is very volatile and toxic to biota and to man.

Apart from the impact of oil on the marine environment, the countries of the Region are also deeply concerned about the introduction of invasive species carried by ballast water and sediments to the RSA. This is particularly noted since the volume of ballast water discharge in our Sea Area is one of the highest in the world and the potential risk of invasion then is a sobering thought. Lack of background information and historical records on the species of our Sea Area is a major hindrance for our investigation of such invasions.

ROPME welcomed the initiative of Global Ballast Water Management Programme – GloBallast in designating the Kharg Island as one of the six global demonstration sites for the study of invasive species carried by ballast water of ships into the Region. ROPME and Member States participated in the IMO Conference on Invasive Species which was held in June 2002 and prepared the text of a Regional Action Plan to minimize the transfer of harmful aquatic organisms and pathogens in ships' ballast water to the RSA. The Regional Action Plan has recently been adopted and countries of the Region are willing to assist IMO in their endeavor on the implementation of the Plan.

ROPME has been supporting every effort towards the ratification of MARPOL Convention and meeting the requirements of adequate reception facilities in the Region in order to declare the RSA as a "Special Area". ROPME shall prescribe the systems, service charging, terms for the operators, and in the meantime, shall provide for monitoring and surveillance of the entire marine environment. The Project is a concerted regional response to the problem of ballast water in the Sea Area since as long as there is the discharge, there will be the threat of invasion by foreign species in addition to oil pollution. Needless to say, the establishment and management of reception facilities will be much facilitated when Port State Control Procedures is adopted and the large number of sub-standard oil tankers and barges are prevented from using the ports of the RSA. Countries of the Region are also carefully monitoring hundreds of war-related wrecks and sunken vessels in the northern part of the RSA for their potential risk of environmental pollution and degradation, and are to study the feasibility of wreck removal in an environmentally sound manner. This of course shall require greater efforts from the concerned UN organizations such as IMO, UNEP, and UNDP.

9. Conclusion

The Kuwait Regional Convention and its Protocols have made a substantial positive impact towards the protection of the marine environment and coastal areas of the Region from oil pollution. However, oil pollution has been and still is a continuous phenomenon in the marine environment and a major challenge for the Region. It should be noted that the exploration, exploitation and transportation of oil has its international elements and as such the international community has the obligation to assist the Region in marine emergencies, and in reducing the impacts of oil exploration, production, and export activities on the Sea Area. In fact, such assistance was provided by the international community to Contracting States directly, or through IMO and UNEP during the 1991 Oil Spill catastrophe. Further technical and scientific supports of the international community are needed to enhance the emergency response and pollution control capabilities of the Region.

Our marine environment occupies a unique position on the surface of the earth and is endowed with valuable mineral resources and a great biodiversity of plant and animal species. But, this fragile and endangered marine ecosystem has received the brunt of many aggressions and catastrophic incidents, and is in urgent need of care. The impacts of millions of barrels of spilled oil during three deteriorating wars and the deployment of chemicals, explosives, and other warfares are enormous. The toll of fish, dolphins, dugongs, whales, waterfowl, algae, corals, and other forms of marine life has set a record. The effects of chronic oil pollution are significant and the specter of oil components passing through the environment and food chain is a sobering thought. Let us join hands to reverse the trend, stop further degradation, salvage resources, and assist nature to rehabilitate. In that direction lies a healthy sea and better lives for our people and the generations to come.

References

Kuwait Regional Convention for Cooperation on the Protection of the Marine Environment from Pollution, 1978.

Protocol concerning Regional Cooperation in Combating Pollution by Oil and other Harmful Substances in Cases of Emergency, 1978.

Kuwait Action Plan for the Protection and Development of the Marine Environment and the Coastal Areas, 1978.

Protocol concerning Marine Pollution resulting from Exploration and Exploitation of the Continental Shelf, 1989.

Protocol for the Protection of the Marine Environment against Pollution from Land-Based Sources, 1990.

Protocol on the Control of Marine Transboundary Movements and Disposal of Hazardous Wastes and Other Wastes, 1998.

Oil Pollution and its Environmental Impact in the Arabian Gulf Region
M. Al-Azab, W. El-Shorbagy and S. Al-Ghais, editors

Chapter 15

Oxidative degradation of oily wastewater by the application of UV-catalytic treatment technologies

Tahir Imran Qureshi[1,*], J.H. Lee[2] and Young-Ju Kim[2]

[1]*Department of Chemical Engineering, NFC-Institute of Engineering and Technological Training, Fertilizer Project, Multan, Pakistan*

[2]*Department of Environmental Engineering, Kyungpook National University, Daegu, South Korea*

Abstract

Combination of different systems of advanced oxidation processes (AOPs) were tested for the treatment of wastewater collected from an oil recycling plant. Oxidation systems were monitored by the global parameters like COD, TOC, VOCs, and UV-light absorbency at 254 nm. Different dosages of oxidants were supplied in a quest finding the optimal conditions to achieve the maximum removal of pollutants from the wastewater. UV-light with the combination of two oxidants, i.e. hydrogen peroxide and ozone proved the highest COD and TOC removal efficiency (>90%). UV/H_2O_2, with higher stoichiometric dosage of peroxide, also succeeded in achieving 70% removal efficiency for the same parameters. Ozone induced degradation system (UV/O_3), even with a high dosage of oxidant, showed the least COD removal efficiency (≈50%), however, the same system showed a little bit higher trend (≈60%) in TOC removal efficiency. Low operating pH conditions favored the efficiency of peroxide induced degradation, while the same might have hindered the efficiency of ozone induced degradation system. Amount of TCE was reduced to 80%, from 5 to 0.02 ppm, with the application of high dosage of peroxide in UV/H_2O_2 system. Comparative statement drawn on the basis of energy consumption sketches suggested that the oxidative degradation pattern of $UV/H_2O_2/O_3$ system was the most cost efficient than the other ones tested.

1. Introduction

Wastewater generated during heat exchanging operation of recycling of exhausted oil bears a high toxic value, of which, major share comes from organic fraction of the pollutants. Since several components of the petroleum products are mutagenic in character (Rubin et al., 1976; Payne et al., 1978), concerns have been raised over possible hazards of their water-soluble substances into the receiving streams. Researchers have been offering technical hands to treat oil-laden wastewater by employing physical and/or chemical

*Corresponding author. *E-mail address:* tiqureshi@yahoo.com (T.I. Qureshi).

means as well since its discharge. Bansal (1976) and Bodzek and Konieczny (1992) considered the oily wastewater treatment using membrane filtration (UF). Results from laboratory pilot plant tests had found such a treatment process efficient and economically attractive but could not reduce COD effectively.

Owing to complex nature of the pollutants, it is not easy to single out the major COD contributors in the wastewater recovers during oil recycling operation. However, it is generally believed that various kinds of aliphatic (acyclic and cyclic), and aromatic hydrocarbons are responsible (Holger and Margrit, 1995). Treatment of oily wastewater by conventional biological or any chemical method is thought to be difficult because of its high VOCs, SS and COD concentrations (Bennett, 1973; Kim et al., 1989). As a matter of fact, refractory character of some pollutants outweighs the utilization of many chemical means and suggests application of destructive type method of high oxidation potential using advanced oxidation processes (AOPs) for their effective degradation (Rice and Cotruvo, 1978).

The synergy of UV-light with oxidants like hydrogen peroxide and ozone for the decomposition of various organic pollutants has been proven to be very effective (Ho, 1986; Sundstrom et al., 1986; Castrantas and Gibilisco, 1990), because hydroxyl radicals (OH) produced possess rather high oxidation potential (Prengle and Mauk, 1978), and can attack on organic fraction by inducing serial oxidation reactions. The hydroxyl radical activates organic compounds by abstracting hydrogen atoms or by adding to double bonds making them to be oxidized more effectively and quickly (Ogata et al., 1981).

Different modes of oxidation in AOPs have been in practice to increase the oxidation level of molecules and to oxidize a larger spectrum of organic compounds. Hydrogen peroxide/ozone (H_2O_2/O_3), and UV/ozone (UV/O_3) systems provide non-selective degradation by free radical oxidation (Hoigne and Bader, 1976). Paillard et al. (1988) stated that oxidation of the pollutants may be accompanied by significant reduction in total organic carbon (TOC) when hydrogen peroxide/ozone is applied at an optimal ratio of 0.35–0.45 g/g in neutral pH. In UV/H_2O_2 process, the optimum H_2O_2 dosage should be obtained and applied since the excess H_2O_2 dosage can reduce the oxidation rate. Results from the experiments done by Juang et al. (1997) revealed that the direct pre-treatment with the UV/H_2O_2 process destroyed the recalcitrant compounds of the petrochemical wastewater into small molecules and might reduce some degree of inhibition to bio-culture. The presence of some metal ions like Cu^{2+} will enhance the H_2O_2 decomposition and increase the rate of organic oxidation, however, the carbonate/bicarbonate ions will inhibit the oxidation of organic fraction due to their scavenging effect (Liao and Gurol, 1995).

This study gives a comparative look on the effectiveness of different AOPs in combination of oxidants with UV-radiation, i.e. UV/H_2O_2, O_3/H_2O_2, UV/O_3, and UV/H_2O_2/O_3 for the treatment of wastewater produced during the heat exchanging operation of an oil recycling plant. Oxidation systems were monitored by the global parameters such as COD, TOC, and absorbency at 254 nm. Aim of this work was to examine influence of UV-light at a minimum intensity, energy requirement of the system and doses of H_2O_2 and O_3 in evaluating applicability of the UV-catalyzed oxidative degradation of the wastewater.

2. Materials and methods

2.1. Equipment

UV-reflecting type reactor was employed for this work with the dimension of 19 cm height, 24 cm length, and 18 cm width. Feed volume 8 l was taken in all the oxidation experiments. A high-pressure UV-lamp consisted of one 20 cm synthetic quartz tube with inner and outer diameters 2.5 and 4.0 cm was installed in upper lid of the reaction vessel and remained 2 cm high above the reaction mixture. Power and specific power of the UV-lamp could be regulated between 300–3000 W and 15–150 W/cm, respectively. Intensity of the light was set at the minimum value 0.69 W/cm^2 by the variable voltage transformer. Input of the UV-light, supplied to the reaction mixture, was 1500 W till the completion of reaction time 3 h. Reflection of the UV-light was facilitated with the help of an arc shaped reflector (24 cm length, 12 cm diameter) adjusted over the UV-lamp. Cooling system of the UV-lamp was maintained with the supply of air at a flow rate of 400 m^3/h. To increase the contact time of ozone with the sample, an ozone contactor (450 cm height, 8 cm diameter) was installed at the inlet of the main reaction vessel and the reaction mixture was recycled at a flow rate of 2 l/min. The column was packed with cylindrical shaped glass media (8 mm height, 4 mm diameter). Ozone was produced from dry oxygen by an ozone generator (COM-CD-HF2, Anseros). Ozone concentration in the gas phase was determined by iodometery titration after trapping it in a potassium iodide solution. All the mechanical parts of the photo reactor assembly contacting either the gas or the liquid phase were made of chemically high resistant materials, such as low carbon stainless steel, perfluorinated plastics or glass. Final treated samples for laboratory analyses were collected from the bottom of the photo reactor. All reactions were carried out at 20 ± 2°C. Figure 1 shows the schematic diagram of UV-reflecting type reactor.

2.2. Sample and operating modes

Samples were collected from the heat exchanger unit of an oil recycling plant located in Yangsan-City. The plant recycles exhausted oil at an operating capacity of 50 m^3/day, and major portion of the used oil is shared by motor/engine oil. Condensate produces in the heat exchanging unit during the final stages of recycling operations at a flow rate 1.35 m^3/day with a COD value ranges 2000~2500 mg/l. Since, the ordinary COD value of the condensate falls in the lower range, i.e. 2000 mg/l, therefore, in this experiment, samples were treated with lower range of COD values. Characteristics of the condensate are given in Table 1. pH of the sample was made acidic from pH 7.5 to 3.5 by the addition of concentrated sulfuric acid.

Batch experiments were carried out and a solution of H_2O_2 (30% w/w) was injected at different stoichiometric amounts equivalent to the amount of COD of the sample (Table 2). Treated samples were taken at the time intervals of 5, 10, 20, 30, 60, 120, and 180 min, the total reaction time. Samples for TOC removal were tested by using TOC analyzer (Shimadzu, TOC-5000A), and TCE was tested by employing GC technique. As the production of hydroxyl radicals from hydrogen peroxide requires large dissociation energy (213 kJ/mole) in order to cleave the O–O bond, it means that short wave UV-C energy

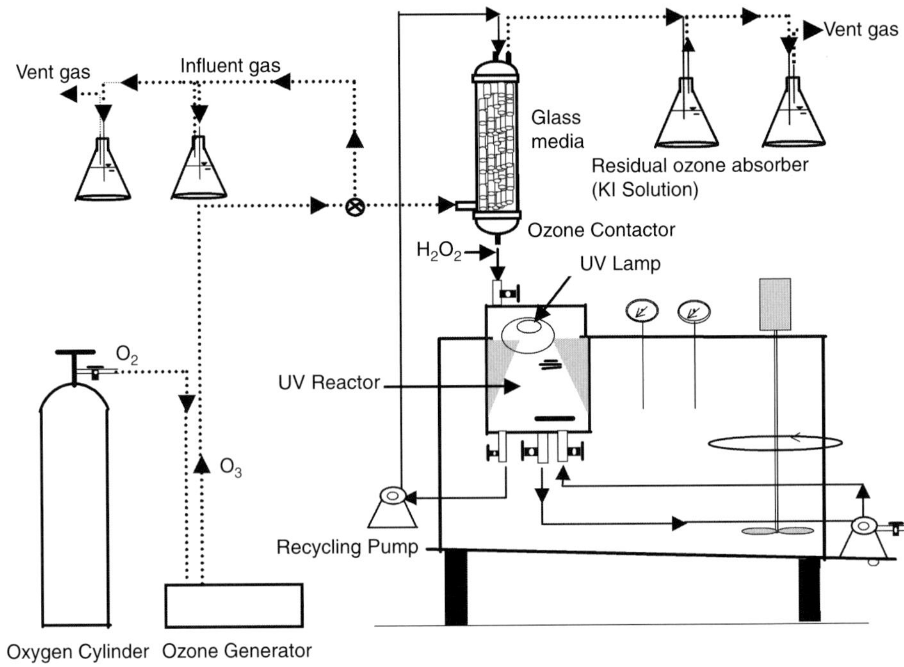

Figure 1. Schematic diagram of UV-reflecting type reactor.

(wavelengths of 200–280 nm) is necessary to achieve useful radical yield (Atkins, 1990). In this study, therefore, high pressure UV-lamp was used to get light of wavelengths 190–350 nm.

3. Results and discussion

3.1. COD removal

All the four systems applied were proven, more or less, successful in reducing COD of the wastewater. The highest COD removal (90%) was recorded in UV/H_2O_2/O_3 system

Table 1. Characteristics of condensate at pH 3.5.

Parameter	Concentration (mg/l)	Parameter	Concentration (mg/l)
COD	2000 ~ 2500	As	0.0117
TOC	760 ~ 780	Hg	ND
TCE	4 ~ 5	Pb	ND
TECE	0.02	Cd	ND
BTEX	ND	Cr^{6+}	ND
n-Hexane	ND	Cu	ND

ND = not detected.

Table 2. Operating conditions in different modes of oxidation experiment.

Oxidation process	Dose of hydrogen peroxide and ozone*
UV/H_2O_2	H_2O_2 (0.2 time)
	H_2O_2 (0.5 time)
	H_2O_2 (1.0 time)
O_3/H_2O_2	O_3 (25 mg/min)/H_2O_2 (0.5 time)
UV/O_3	O_3 (25 mg/min)
	O_3 (50 mg/min)
$UV/H_2O_2/O_3$	H_2O_2 (0.5 time)/O_3 (25 mg/min)
	H_2O_2 (1.0 time)/O_3 (25 mg/min)

* Stoichiometric amount equivalent to the amount of COI of the sample.

(Fig. 2). The UV/O_3 system, which is regarded as ozone induced degradation, manifested the least COD removal (50%). The synergy of two oxidants with UV-light increased the oxidation potential to an extent that the COD removal efficiency up to 85% was achieved within 2 h of reaction time (Fig. 2). While the other two systems, i.e. UV/H_2O_2, and UV/O_3 could not obtain the same degradation potential even with an extended reaction time 3 h. The cost involved with the process UV-light with two oxidants could be lower than the cost for a system utilizing only UV-light and ozone, since addition of H_2O_2 allowed use of a smaller ozone generator with less oxidant.

UV/H_2O_2 was also succeeded in obtaining potential removal of COD, up to 71% (Fig. 3), which is 40% higher than the value obtained form the ozone induced degradation system (Fig. 4). Hydrogen peroxide with the combination of UV is regarded as peroxide induced degradation and the system is favored at low pH values. Since rate of oxidation of organic compound depends on the relative concentration of H_2O_2 species (H_2O_2 and HO_2^-), therefore, increased pH of the solution would increase both, the generation of hydroxyl radicals to a lesser extent, while the scavenging effects of H_2O_2 to an higher extent, ultimately could bringing the efficiency down. As UV/H_2O_2 system was operated at low pH, therefore, high COD removal was observed in comparison to the ozone induced degradation systems.

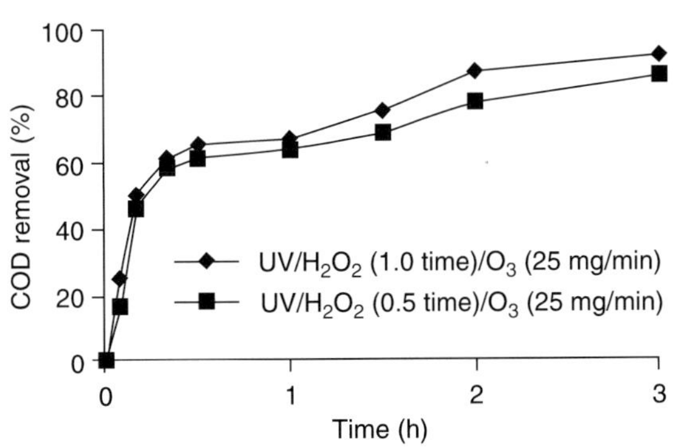

Figure 2. Profile of COD removal in different modes of $UV/H_2O_2/O_3$ system.

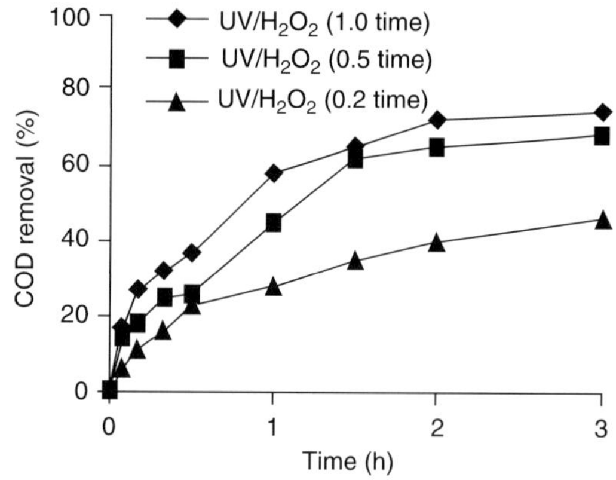

Figure 3. Profile of COD removal in different modes of UV/H_2O_2 system.

Ozone with UV-light (Fig. 4) or with hydrogen peroxide, UV/H_2O_2 (data are not shown) could barely achieve the COD removal efficiency to 50%, even with higher doses, till the completion of total reaction time, 180 min. This could be explained on the basis that ozone induced degradation is favored at high pH values and our system was operated at low pH conditions.

High pH refers to the presence of a high concentration of hydroxyl ions ($\bar{O}H$) that may lead to hydroxyl radical formation (OH), the main oxidizing character, through an indirect route. Also, the rate of decomposition of ozone increases with increasing pH, consequently, beyond some critical pH, hydrogen radical could become the predominant oxidizing species. This phenomenon could lead the system to higher COD removal trend provided that the system was operated at higher pH conditions.

3.2. TOC removal

TOC removal experiments were conducted out on the same pattern followed in the case of COD removal. It is evident from the data that the highest TOC removal efficiency, up to

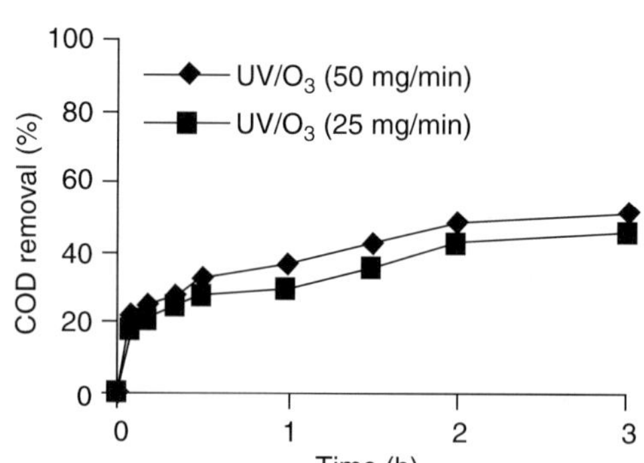

Figure 4. Profile of COD removal in different modes of UV/O_3 system.

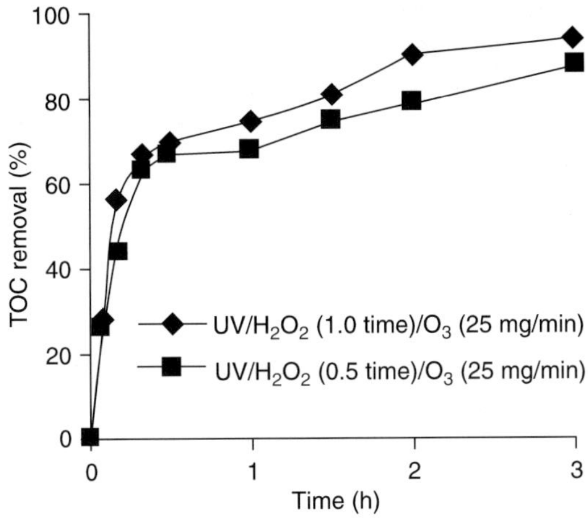

Figure 5. Profile of TOC removal in different modes of $UV/H_2O_2/O_3$ system.

90%, was recorded in the system UV-light with the combination of two oxidants, i.e. $UV/H_2O_2/O_3$ (Fig. 5). Irradiated peroxide system was also succeeded in obtaining significant TOC reduction (75%) within the same interval of time (Fig. 6). The other two systems, i.e. UV/O_3 (Fig. 7), and H_2O_2/O_3 (data not shown), could barely bring the TOC abatement to the level of 60%.

Higher efficiency of the $UV/H_2O_2/O_3$ system could be attributed to the fact that, synergy of UV with two oxidants increases the oxidation level to a strength which would be unlikely to exist with the combination of peroxide or ozone alone. Under the influence

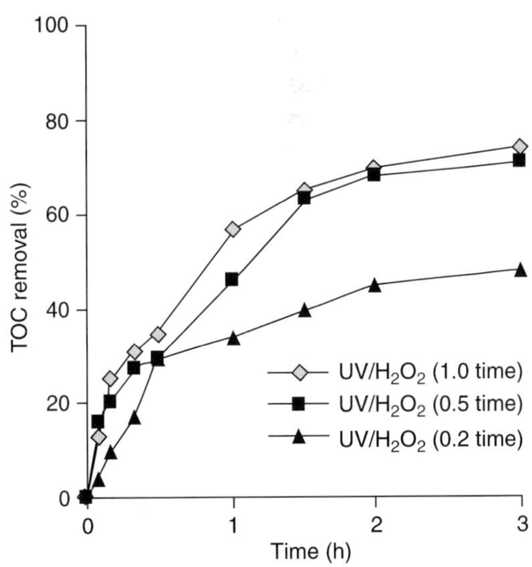

Figure 6. Profile of TOC removal in different modes of UV/H_2O_2 system.

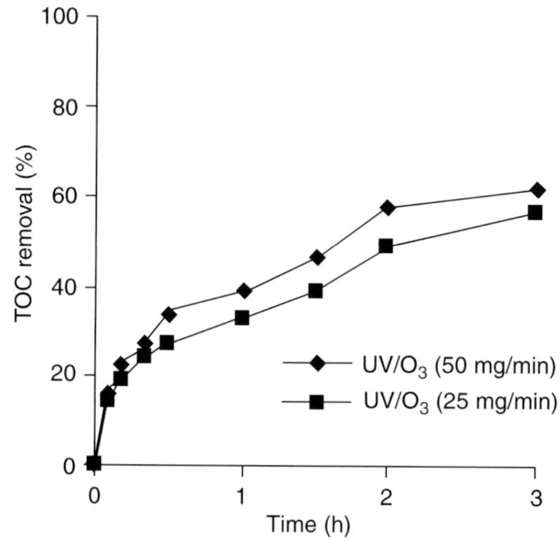

Figure 7. Profile of TOC removal in different modes of UV/O_3 system.

of UV-light, like ozone, H_2O_2/O_3 reacts with aromatic compounds, moreover, it also reacts on aliphatic acids with hydroxyl radicals. These conditions make mineralization relatively higher than the other AOPs.

The least TOC removal (47%) was recorded in O_3/H_2O_2 system (data are not shown). In the absence of UV-light, the formation rate of hydroxyl radicals at acidic pH is low, however, the hydroxyl radicals produced by ozone continue to oxidize the reaction substances into the final products with a relatively slow rate. Since our experiment was conducted at acidic pH, the above mentioned factor might play a role in making the ozone peroxide system less efficient.

However, ozone induced degradation showed a little bit higher trend in TOC removal efficiency (Fig. 7) as compared to the COD removal (Fig. 4). TOC removal was recorded 20% more than the COD removal in respective oxidation systems. This could be addressed in the light of the other investigators' work, which had shown that ozonization was highly effective in breaking down the straight, unsaturated bonds in organic molecules, but regarding COD removal, ozone was considerably less efficient (Lin and Chen, 1997).

3.3. VOCs removal

In case of VOCs, only TCE was found in the sample high above the effluent discharge standard, and other VOCs were present below the detection limit. Therefore, samples were tested only for TCE reduction. The UV/H_2O_2 system with high dose of hydrogen peroxide was succeeded in obtaining 80% removal of TCE within 2 h of reaction time. As the system was succeeded in bringing down the TCE, from 5 to 0.02 ppm, an amount below the discharge limit, therefore, other system, i.e. UV/H_2O_2/O_3 was not applied for this test.

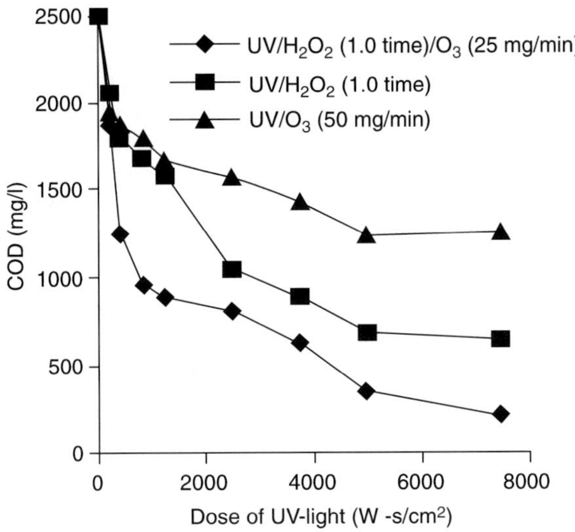

Figure 8. Sketch of energy consumption requirement regarding COD degradation potential in different oxidative systems. Specific power of the lamp could be regulated at 15–150 W/cm while intensity and input of the UV-light were set at 0.69 W/cm^2, and 1500 W, respectively till the completion of reaction time 3 h.

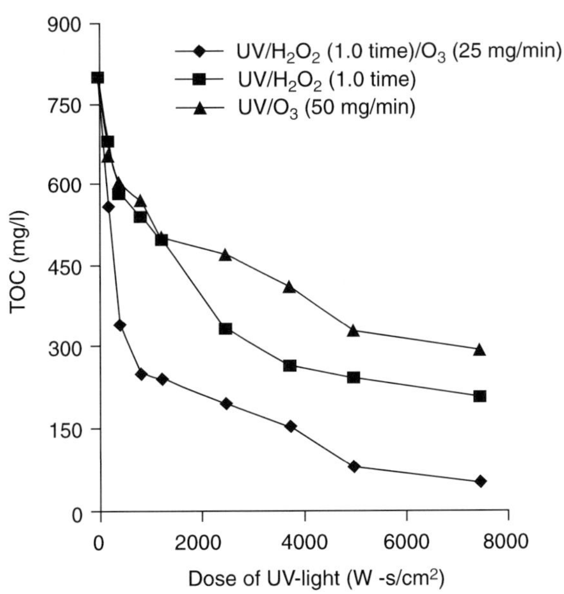

Figure 9. Sketch of energy consumption requirement regarding TOC degradation potential in different oxidative systems. Specific power of the lamp could be regulated at 15–150 W/cm while intensity and input of the UV-light were set at 0.69 W/cm^2, and 1500 W, respectively, till the completion of reaction time 3 h.

3.4. Energy consumption requirement

Energy consumption sketches drawn on the basis of amount of electrical energy utilized versus degradation potential suggested economic preference of one system over the other ones. Intensity of the UV-light was kept at a minimum value (0.69 W/cm^2) throughout the experiment in order to get the highest possible degradation values with the utilization of minimum amount of energy. In term of UV-light dosage supplied to different oxidative systems, $UV/H_2O_2/O_3$ was found the most cost-effective system as compared to the other two systems (Fig. 8). A dose of UV-light (7452 Ws/cm^2) successfully achieved 90% COD removal in $UV/H_2O_2/O_3$ system while the other two systems with one oxidant could not compete in securing similar removal efficiency with the supply of equal dosage of UV-light. Energy consumption requirement for TOC removal followed the same trends as with that of COD removal in all the three oxidative systems (Fig. 9).

4. Conclusions

Four different systems of AOPs' were applied for the treatment of condensate recovered form heat exchanging unit of an oil recycling plant. Synergy of the UV-light with two oxidants, hydrogen peroxide and ozone, was succeeded in achieving about 90% removal of COD and TOC in the respective systems. UV/H_2O_2 system also led the potential removal of COD and TOC to 70%, possibly taking the advantage of low operating pH value. Amount of TCE was reduced to 80%, from 5 to 0.02 ppm, with the application of high dose of peroxide in UV/H_2O_2 system. Ozone induced degradation systems could barely reduce COD below 50% even with the higher dosage of ozone. Acidic condition, prevailed in the reaction mixture might have hindered their efficiency. In contrast to the COD removal, higher TOC removal trend in UV/O_3, could be linked to the effectiveness of ozone in TOC destruction over the COD abatement. Energy consumption sketches also suggested economic preference of $UV/H_2O_2/O_3$ system over the other oxidative systems.

5. Nomenclature

AOPs	advanced oxidation processes
BTEX	benzene, toluene, ethylene benzene, and xylene
COD	chemical oxygen demand
GC	gas chromatography
SS	suspended solid
TCE	tri-chloroethylene
TECE	tetra-chloroethylene
TOC	total organic carbon
UF	ultra filtration
UV	ultraviolet light
VOCs	volatile organic compounds

References

Atkins, P.W., 1990. Physical Chemistry, 4th edn. Oxford University Press, Oxford, p. 938.

Bansal, I.K., 1976. Concentration of oily and latex wastewater using ultrafiltration inorganic membranes. Ind. Water Eng. 13 (10), 6.

Bennett, E.O., 1973. The disposal of metal cutting fluids. Lub. Eng. 29, 300.

Bodzek, M., Konieczny, K., 1992. The use of ultrafiltration membranes made of various polymers in the treatment of oil-emulsion wastewaters. Waste Manage. 15, 75.

Castrantas, H.M., Gibilisco, R.D., 1990. UV destruction of phenolic compounds under alkaline conditions. ACS Symp. Ser. 422, 77.

Ho, P.C., 1986. Photooxidation of 2,4-dinitrotoluene in aqueous solution in the presence of hydrogen peroxide. Environ. Sci. Technol. 20, 260.

Hoigne, J., Bader, H., 1976. The role of hydroxyl radical reactions in ozonation processes in aqueous solution. Water Res. 10, 377.

Holger, G., Margrit, R., 1995. Organic compounds at different stages of a refinery wastewater treatment plant. Water Sci. Technol. 32 (7), 119–126.

Juang, L.C., Tseng, D.H., Yang, S.C., 1997. Treatment of petrochemical wastewater by UV/H_2O_2 photodecomposed system. Water Sci. Technol. 36 (12), 357–365.

Kim, B.R., Matz, M.J., Lapari, F., 1989. Treatment of metal-cutting fluid wastewater using an aerobic GAC fluidized-bed reactor. J. Water Pollut. Control Fed. 8, 1430.

Liao, C.H., Gurol, M.D., 1995. Chemical oxidation by photolytic decomposition of hydrogen peroxide. Environ. Sci. Technol. 29 (12), 3007.

Lin, S.H., Chen, M.L., 1997. Combined ozonation and ion exchange treatments of textile wastewater effluents. J. Env. Sci. Health A2 (7), 1999, part A.

Ogata, Y., Tomizawa, K., Takagi, K., 1981. Photooxidation of formic, acetic and propionic acids with aqueous hydrogen peroxide. Can. J. Chem. 59, 14.

Paillard, H., Brunet, R., Dore, M., 1988. Conditions optimales d' application du systeme oxydant ozone-per-oxyde d'hydrogene. Water Res. 22, 91.

Payne, J.F., Martins, I., Rahimtula, A., 1978. Crankcase oils: are they a major mutagenic burden in the aquatic environment? Science. 200, 329–330.

Prengle, H.W., Mauk, C.E., 1978. New technology: ozone/UV chemical oxidation wastewater process for metal complexes, organic species and disinfection. AIChE Symp. Ser. 74, 228.

Rice, R.G., Cotruvo, J.A., 1978. Ozone/chlorine dioxide, oxidation products of organic materials. Ozone press international, Cleveland, OH, 490.

Rubin, I.B., Guerin, M.R., Hardigree, A.A., Epler, J.L., 1976. Fractionation of synthetic crude oils from coal for biological testing. Environ. Res. 12, 358–365.

Sundstrom, D.W., Klei, H.E., Reidy, D.J., Weir, B.A., 1986. Destruction of halogenated aliphatics by ultraviolet catalyzed oxidation with hydrogen peroxide. Hazard. Waste Hazard. Mater. 3, 101.

Index